The Legacy of
MILTON AND ROSE FRIEDMAN'S
Free to Choose

The Legacy of
MILTON AND ROSE FRIEDMAN'S
Free to Choose
Economic Liberalism at the Turn of the 21st Century

Proceedings of a Conference Sponsored by the
Federal Reserve Bank of Dallas
October 2003

Edited by
Mark A. Wynne
Harvey Rosenblum
Robert L. Formaini

With a Foreword by
Milton and Rose Friedman

Federal Reserve Bank of Dallas
Dallas, Texas

Proceedings of a conference sponsored by the
Federal Reserve Bank of Dallas, October 23–24, 2003
Published December 2004

ISBN 0-9763494-1-8

Articles may be reprinted on the condition that the source is credited and a copy is provided to the Research Department, Federal Reserve Bank of Dallas, P.O. Box 655906, Dallas, TX 75265-5906. Additional copies of this publication can be obtained by calling 214-922-5254. This publication is also available on the Internet at www.dallasfed.org.

A companion video titled *Ideas Matter* featuring interviews with conference presenters as well as scenes from the original television series *Free to Choose* is also available from the Federal Reserve Bank of Dallas.

The views expressed in this volume are those of the authors and should not be attributed to the Federal Reserve Bank of Dallas or the Federal Reserve System.

This book is printed on acid-free paper.

Contents

Foreword
    Milton and Rose Friedman vii

Preface
    Robert D. McTeer, Jr. ix

Remarks
    Alan Greenspan xi

Introduction
"Free to Choose" 3
    Mark A. Wynne

Session 1
The Toughest Battleground: Schools 21
    Eric A. Hanushek

The Theory and Practice of School Choice 37
    Paul E. Peterson

Session 2
The Property Rights Path to Sustainable Development 57
    Terry L. Anderson and Laura E. Huggins

Economic Freedom and Environmental Quality 73
    Richard L. Stroup

Session 3
The Economic Burden of Taxation 93
    William A. Niskanen

The Transition to Private Market Provision of Elderly Entitlements 99
    Liqun Liu, Andrew J. Rettenmaier, and Thomas R. Saving

Session 4
Commerce, Culture and Diversity: Some Friedmanesque Themes in Trade and the Arts 123
    Tyler Cowen

Milton and Rose Friedman's "Free to Choose" and Its Impact in the Global Movement Toward Free Market Policy: 1979–2003 137
    Peter J. Boettke

Free to Choose in China 153
    Gregory C. Chow

Session 5
Financial Markets and Economic Freedom 175
    Luigi Zingales

Choosing Freely: The Friedmans' Influence on Economic and Social Policy 191
    Allan H. Meltzer

Friedman's Monetary Framework: Some Lessons 207
    Ben S. Bernanke

Session 6
What Have We Learned from the Measurement of Economic Freedom? 217
    James Gwartney and Robert Lawson

Can the Tide Turn? 239
    Raghuram G. Rajan

Acknowledgments 249

About the Contributors 251

Foreword

In our 1998 memoir *Two Lucky People*, we described our participation in the filming of the *Free to Choose* television program as "the most exciting venture of our lives." As Rose wrote: "It all seems like something of a fairy tale. Who would have thought that after retiring from teaching, Milton would be able to preach the doctrine of human freedom to many millions of people around the globe through television, millions more through our book based on the television program, and countless others through videocassettes."

And who would have thought that nearly a quarter century later, the program and the book would have enough life to justify the splendid conference that the Federal Reserve Bank of Dallas staged and to bring forth the excellent set of papers reprinted in this volume.

The program and the book are still alive because the problems they deal with—free trade, regulation, business cycles, welfare state, inequality, consumer protection, worker protection, inflation—are all still with us, and how we handle them will affect both our prosperity and our freedom. One very important thing, however, *is* different. Thanks largely to the collapse of the Soviet Union, there has been a dramatic change in the climate of opinion around the world.

A quarter century ago, most people believed that socialism was a viable system for promoting material prosperity and human freedom; many believed it was the most promising system for doing so. Few people anywhere in the world believe that today. Idealistic faith in socialism still lives on, but only in some ivory tower enclaves in the West and in some of the most backward countries elsewhere. A quarter century ago, many people were convinced that capitalism, based on free private markets, was a deeply flawed system that was not capable of achieving both widely shared prosperity and human freedom. Today it is increasingly recognized that capitalism is the only system that can do so.

While the main thesis of *Free to Choose* has become conventional wisdom, conventional practice, at least in the West, has not changed. Political leaders in capitalist countries who cheer the collapse of socialism in other countries continue to favor socialist solutions in their own. They know the words, but they have not learned the tune. The widespread vested interests created by socialist measures retain their political power and continue to resist any major reform. There is a true tyranny of the status quo. And yet, the change in opinion has kept down the growth of government and is widening the role of markets.

The most dramatic changes in the past quarter century have come in the formerly communist countries. In those, literally billions of people have come from under despotism and achieved a greater measure of freedom. For many, the freedom is still highly limited—greater economic freedom but still limited civic and political freedom. However, the transition is still in its early stages and cannot be stopped. The change is unprecedented.

The essays in this book deal with the same subjects as the TV program and the book *Free to Choose*, but they do so in tune with today's conventional wisdom, not that of a quarter of a century ago. We found them refreshing and informative to hear and to read, and I am sure you will as well.

We are very much indebted to Robert McTeer, Jr., president of the Federal Reserve Bank of Dallas, and to his associates, especially Harvey Rosenblum, Mark Wynne, and Robert Formaini, for organizing this splendid conference and persuading so many eminent scholars to present papers. We especially appreciate the effort they took to make the experience pleasurable for two nonagenarians.

MILTON AND ROSE D. FRIEDMAN

Preface

In an era of economic transition, rapid technological change, and globalization, Milton Friedman's message of economic freedom is more compelling and relevant than ever. His famous maxim about the impossibility of free lunches reminds us that there are costs and trade-offs in everything we do, and we should look at what the alternatives are and who picks up the tab.

Friedman recognizes the power of the invisible hand of free enterprise to create wealth and jobs, while warning that the heavy hand of government will bring nothing but stagnation. He has argued for a monetary policy to stabilize prices and keep inflation low.

Most important, Friedman has made economics a moral matter as well as one of productivity, jobs, and growth. Economic freedom, he reminds us, is every bit as precious as the other freedoms we treasure.

The papers in this volume were presented at a conference the Federal Reserve Bank of Dallas hosted in 2003 to pay homage to the Nobel laureate's life and work. We were fortunate that both Milton and Rose Friedman were able to attend and participate in all the sessions. The quality of the speakers and their contributions speaks to the high regard in which the Friedmans are held by all who believe in economic freedom.

ROBERT D. MCTEER, JR.
President and Chief Executive Officer
Federal Reserve Bank of Dallas

Remarks

Ideas matter. Indeed, the world is ruled by little else. Many years ago, John Maynard Keynes wrote that "practical men, who believe themselves to be quite exempt from intellectual influences, are usually the slaves of some defunct economist. Madmen in authority, who hear voices in the air, are distilling their frenzy from some academic scribbler of a few years back."

Milton and Rose are indeed academic scribblers, though hardly defunct, and not surprisingly, their views have become pervasive throughout this culture. The radical notion of freedom espoused by them has echoed down the generations to create societies where individuals acting in their own self-interest have, by engaging in mutually beneficial exchange under a rule of law, elevated standards of living for a rising share of the world's population.

For more than a half century, Milton and Rose Friedman have been in the vanguard of removing the stultifying weight of the state, rendering an ever larger part of the human race free to choose. They move their world by a combination of scrupulous adherence to fact and an extraordinarily efficient thinking process. In fact, this process has fostered what some of Milton's detractors—there are indeed some—believe is a put-down but is in fact an inadvertent compliment. The line goes, "I wish I was as certain about *anything* as Milton is about *everything*." He is certain about what he knows, but shouldn't everybody?

Professor Friedman is, as you all know, a formidable debater, a characteristic in stark evidence three decades ago during, as Milton would remember, a session of the commission on an all-volunteer armed force, on which he and I served. I do not recall the details except that in a very quiet way Milton dismembered a very famous general's position in favor of the draft and against what the general termed a mercenary armed force. I later spoke to Tom Gates, who chaired the commission, who said that he too had opposed eliminating the draft but Milton turned him around. That may not seem significant except he was a former secretary of defense.

Milton's certainty of who he is has led to another admirable characteristic. He, as all of you are acutely aware, is utterly without guile or proclivity to spin or manipulate. He does not speak down to an audience nor alter his content to gain their approval. He presumes that he is speaking to a rational, if not always knowledgeable, audience. In discussions on issues, he exhibits the same demeanor whether he is talking to a high school student or the president of the United States.

It is this characteristic that once helped me frame a response to requests to televise Federal Open Market Committee meetings. I said most people unduly hedge their opinions if they are recorded, especially live. I thought it would alter the nature of what FOMC members said and undercut freewheeling debate. If you could fill the FOMC with Milton Friedmans, I acknowledged, I would have no such hesitation with live cameras. I do not ever recall Professor Friedman's views being altered by television cameras or, for that matter, by anything else.

My first contact with Milton was in 1959, when I mailed him a copy of an article on the impact of the ratio of stock prices to replacement cost on capital investment. I am sure he had never heard of me, yet he took the time to reply with several very thoughtful suggestions. I have never forgotten that.

I find I cannot discuss Milton without discussing Rose. I have never been able to visualize Milton without Rose in the background inspiring and correcting him. As I indicated in a White House tribute to Milton early last year, I'm not certain who is the more intellectually formidable of the two. Perhaps that owes to the fact that I was stirred first by Rose Friedman's insights before I really fully understood Milton's contribution to economic thought.

In 1949, I had started on a doctoral thesis on the savings rate under Arthur Burns of Columbia. I soon ran into a rather intriguing article by Dorothy Brady and Rose Friedman that demonstrated, if my memory serves me, that a particular household savings rate is a function of that household's income relative to the average of the total relevant population. It explained why the savings rate has no trend through time despite the presumption that the higher the income level, the higher the savings rate. As a 23-year-old, I was utterly fascinated by that insight.

Other participants in today's program will, I trust, present the many accomplishments of both Rose and Milton to the intellectual heritage of the world and the profoundly influential role they have played in furthering human freedom. I wish to add that perhaps their most important accomplishment is the creation of themselves—in their language, two lucky people. If I believed in luck, I would argue that the most lucky people are those of us who have had the privilege of knowing both of them.

ALAN GREENSPAN
Chairman
Federal Reserve Board

Introduction

"Free to Choose"

Mark A. Wynne

"Free to Choose"

Mark A. Wynne

In 1962 Milton Friedman published *Capitalism and Freedom*, one of the most influential arguments for economic liberalism to appear in the second half of the twentieth century. *Capitalism and Freedom* has been in print for the past forty years and has been translated into no fewer than eighteen languages. At the time it was published, the book was not widely reviewed outside of the main academic journals.[1] But with the book's publication, Milton Friedman staked his claim as a champion of economic liberalism at a time when the ideas of liberals (in the traditional sense) were distinctly unfashionable.[2]

Some twenty years later, in 1980, Milton and Rose Friedman made the case for economic freedom to a broader audience with the PBS television series and book *Free to Choose*. This book was highly successful, becoming the best-selling nonfiction book of 1980, and presented their agenda for reform. In the preface to the 1990 (Harvest) edition of *Free to Choose*, the Friedmans wrote of how surprised they were at the dramatic turning of the tide that had occurred in the 1980s. They were prompted to wonder whether the ideas in *Free to Choose* had become so much part of the conventional wisdom that the book was no longer relevant.

In 2003 the Federal Reserve Bank of Dallas organized a conference to take a retrospective look at *Free to Choose*. A large part of the motivation for holding the conference, over and above the desire to honor one of the twentieth century's greatest American economists, was a concern that the United States and the world were possibly at another turning point, but this time away from small government and freer markets. Protesters against globalization have become increasingly strident in their denunciations of market capitalism and free trade. In some regions of the world, voters are asking whether market reforms of recent years have paid off. And the trend toward freer trade seems to have stalled as governments have chosen to put narrow domestic interests ahead of principle.

This essay provides an overview of some of the issues discussed in the papers that follow. I begin with a summary of some of the central ideas in *Capitalism and Freedom* and *Free to Choose*. Next I examine how the world looked in 1980 and how it has changed since then, paying particular attention to areas where the ideas expressed in *Free to Choose* have been influential. I then consider some threats to economic freedom that may lead to a rolling back of some of the advances of the past two decades.

THE BASIC MESSAGE OF *CAPITALISM AND FREEDOM* AND *FREE TO CHOOSE*

Free to Choose is probably best summarized by this statement toward the end of the book: "Reliance on the freedom of people to control their own lives in accordance with their own values is the surest way to achieve the full potential of a great society" (Friedman and Friedman 1980, 309–10). In the economic sphere, this means relying on free private markets as the primary means of organizing production and exchange, with a minimal role for government. At a time when many were still in the thrall of state planning, the Friedmans took the distinctly unfashionable stance of arguing for minimal government involvement in the economy.

Both books make three key points. First, free competitive markets are the most effective way to organize production and exchange and to ensure that the wants of the people are met. The power of competitive markets to deliver desirable outcomes was Adam Smith's great insight and remains as relevant today as it was when first articulated in 1776. Second, when the government intervenes to rectify a case of market failure, often the cure is worse than the disease. Many of the so-called failures of capitalism, especially the Great Depression of the 1930s, were due to misguided government policies rather than inherent weaknesses in the capitalist system. Third, free markets in conjunction with equality of opportunity allow individuals to attain standards of living previously thought unattainable. The gap between the rich and the poor tends to be greatest in societies where the free market is suppressed. Putting equality ahead of freedom will cost a society both; putting freedom ahead of equality is the surest guarantor of both.

The opening chapter of *Free to Choose*, titled "The Power of the Market," provides the basic framework used to address a variety of issues. Competitive free markets consistently deliver what consumers want, at lower cost, than any other mechanism known to man. This is true whether the market is for breakfast cereal, cars or educational services. In his contribution to this volume, **Paul Peterson** reviews the evidence on school choice and shows that, along almost every dimension, schools are better at delivering what parents want when there is an element of competition in the provision of education. The exact form of competition—school vouchers or charter schools—is less important than the presence of competition. Markets for education work just as well as markets for agricultural commodities or foreign exchange. Speaking at the conference dinner, **Gary Becker** reiterated the key point of the power of competition.[3] Arguing that competition is probably the most important social contrivance of the last thousand years, Becker pointed out the key characteristics of competition: It drives down costs; it fosters innovation; it drives up quality; and most important, it economizes on information. Just as competition displays these characteristics when allowed to work in the market for consumer goods, so too will it lower costs, foster innovation, improve school quality and economize on information if allowed to work in primary and secondary education. Becker noted that it was no accident that the United States has the best third-level education system in the world, attributing this to the greater degree of competition in this segment of the education system.

The potential of free markets to raise living standards is only realized when individuals are free to specialize in doing what they do best and trade for their other needs. Trade through the medium of money is most efficient, and fiat monetary standards economize on the resource costs of monetary exchange. But fiat monetary standards come at a cost, that of inflation.[4] Through the 1970s and 1980s, inflation accelerated to rates that had not been seen in many countries. Friedman argued early that the government need only set some predetermined growth rate for the stock of money, thereby eliminating all discretion from the conduct of monetary policy, to control inflation and the real instability associated with discretionary monetary policy. In recent years, Friedman has come to accept that his preferred policy prescription of strict monetary targeting would not have worked very well if it had been widely implemented, and **Ben Bernanke** notes in his paper that this is the only part of Friedman's monetary framework that has not become part of the contemporary conventional wisdom on best practices in monetary policy. But Friedman has also noted that many central banks seem to have adopted his key policy prescription (and the central message of chapter 9 of *Free to Choose*) that control of the money stock is the key to control of inflation.[5]

Bernanke states that Milton Friedman's monetary framework "has been so influential that…it has nearly become identical with modern monetary theory and practice." One of Friedman's key insights was that while money may influence real activity in the short run, it has no effect in the long run. Monetary policymakers' failure to appreciate that insight contributed to the Great Inflation of the 1970s, which Bernanke describes as the second great monetary mistake of the twentieth century. The first, of course, was the Great Depression. In chapter 3 of *Free to Choose*, the Friedmans examine the Great Depression and restate the argument first developed in Friedman and Schwartz (1963) that the Depression was fundamentally due to errors on the part of the Federal Reserve System.

Bad monetary policy turned what otherwise would have been a run-of-the-mill recession into a major depression.[6] As testimony to how well the current Federal Reserve System has learned this lesson, in a panel discussion on the Great Depression at a University of Chicago event held in 2002 to mark Friedman's ninetieth birthday, Bernanke concluded with the confession: "You're right, we did it. We're very sorry. But thanks to you, we won't do it again."

The Friedmans argue that the greatest threat to economic freedom comes from the government. Government intervention in the economy comes in many forms, from regulation of some economic activities to prohibition of others, to monopolization of yet others and appropriation of resources through taxes and other levies. One of the most pernicious such levies in the United States for a long time was the draft of young men into the military, which Friedman campaigned against vigorously until its repeal in 1973. The draft is mentioned in chapter 2 of *Capitalism and Freedom* as one of fourteen activities undertaken by the U.S. government that was inconsistent with liberal economics. By the time *Free to Choose* was written, the draft had been abolished, due in no small part to the efforts of the Friedmans, but the presence or absence of conscription is one of the key components of the economic freedom indexes that have been developed in response to the Friedmans' work. **James Gwartney** pioneered the construction of indexes of economic freedom, and in his paper he documents the tight relationship between economic freedom and economic growth. Gwartney shows that those countries that maintain institutions and policies consistent with greater economic freedom also tend to have higher per capita GDP. Economic freedom enhances productivity both directly and indirectly by boosting investment. Gwartney finds that increases in economic freedom during the 1980s seem to have a statistically significant positive effect on long-run growth: specifically, a one-unit increase in the index of economic freedom during the 1980s enhanced long-term growth by 0.71 percentage point.

In *Capitalism and Freedom*, Friedman devoted a chapter to exploring the relationship between economic and political freedom. As interest in this relationship has developed in subsequent years, it has become apparent that a third category of freedom needs to be added to the mix, namely civil freedom. Friedman himself has argued this need in a number of venues in recent years and again at the Dallas conference. Hong Kong under British rule was the prime example of a society that enjoyed a high degree of economic freedom and civil freedom (freedom of speech and freedom of association), but limited political freedom: The colony was essentially run as a benevolent dictatorship by the British Foreign and Commonwealth Office. A major challenge for the economic freedom project, in Friedman's view, will be to integrate measures of economic freedom with measures of political freedom and reconcile the two where they conflict.

When *Capitalism and Freedom* and *Free to Choose* were written, equality was one of the thorniest issues the Friedmans grappled with. It remains a difficult issue today. As the Friedmans note in *Free to Choose*, "A society that puts equality—in the sense of equality of outcomes—ahead of freedom will end up with neither equality nor freedom....[but] a society that puts freedom first will, as a happy by-product, end up with both greater freedom and greater equality" (148). Equality of opportunity is not directly addressed in any of the conference contributions but runs through many of them as a leitmotif. Perhaps the easiest way to improve equality of opportunity in the United States would be to promote competition in the K–12 education system, which would improve school quality and the range of educational opportunities available to all children. In his paper, **Raghurum Rajan** notes that elites in many societies tend to undermine equality of opportunity by opposing widespread access to markets, often by limiting access to finance. One of the keys to ensuring the political viability of free markets and the greater opportunities they create for all is to get the elites behind markets.

But free markets come with important prerequisites. In his paper, **Luigi Zingales** also emphasizes the importance of access to finance in allowing individuals to realize their full potential under the capitalist system. Zingales starts with the story of Sufiya Begum, a stool maker in an impoverished Bangladeshi village, to illustrate how a lack of access to finance can hinder the ability of individuals to advance even with free markets. For want of access to finance, Begum is effectively indentured to a single middleman who exploits his position of monopoly and monopsony power to limit her income. Zingales argues that access to finance is crucial to promoting competition and ensuring maximum economic freedom. An important corollary is that legislation limiting access to finance, whether intentionally or not, can have detrimental effects on the ability of individuals to realize the full benefits of free markets.

Critics of free market capitalism like to dismiss it as being too concerned with material things and detrimental to the development of culture. In 1993, the prime minister of France, Eduard Balladur, asked rhetorically, "What is the market? It is the law of the jungle, the law of nature. And what is civilization? It is the struggle against nature." This summarizes the view of many critics of free market capitalism, especially European critics. Many of these critics believe that capitalism is detrimental to the finer things in life. Yet as **Tyler Cowen** argues in his paper, the wealth and freedom that capitalism makes possible are a boon for the arts. Indeed, periods of greater globalization also tend to be periods of greater cultural diversity and creativity. The greater wealth that capitalist societies generate supports a greater range of cultural products and makes it easier to preserve the pasts of their own and other societies. While it may be too early to say how much of the art produced in twentieth century capitalist societies will be deemed great by future generations, it is noteworthy that much of the art that has survived the test of time was funded by private patrons living in wealthy societies. Cowen cites Renaissance Italy, the Dutch Golden Age, and the blossoming of French culture in the nineteenth century as examples. The antiglobalization protestors who pose such a threat to the liberal economic order rail against the McDonaldization of the world. They see the spread of American culture overseas but overlook the spread of foreign culture to the United States. As Cowen acknowledges, free trade may indeed reduce diversity across societies, but it invariably increases diversity within societies.

THE WORLD IN 1980

How did the world look in 1980 when *Free to Choose* was published? Government was large, the Cold War was at its height and the Friedmans' ideas were still far removed from the mainstream. In the United States, total government expenditures at the federal, state, and local levels accounted for 30 percent of GDP. In other countries, the extent of government involvement in the economy was even greater—large sectors of the economies of the major Western industrial countries were under direct government control. In the United Kingdom, the steel industry, railways, coal mining, and a myriad of other sectors had been nationalized by various postwar governments. The Mitterand government that came to power in France in 1980 marked the last great push for greater government control in a major Western industrial economy.

Much of the high level of U.S. federal government expenditure was devoted to defense. The United States had some half million troops stationed in Western Europe, where it was engaged in a face-off with the Warsaw Pact countries. The Iron Curtain was tightly drawn across the face of Europe. Indeed, military spending was on the increase—the United States was in the process of stationing cruise missiles in Western Europe to counter the Soviet military threat, to much public (European) opposition. In 1979 the Soviet Union had invaded Afghanistan, the latest in a series of Soviet interventions during the postwar era to retain its grip on its satellite states (beginning with East Germany, Hungary, and Czechoslovakia). To many, the Soviet Union and its allies looked invincible.

Around the world, inflation was running at levels not seen since the end of World War II. In the United States, consumer price inflation was 13.5 percent in 1980, its highest level since 1947. Elsewhere in the industrial world, inflation was at or near the highs experienced during the 1970s. In many countries, wage and price controls remained the preferred approach to dealing with inflation. In the eleventh edition of his influential textbook *Economics*, published in 1980, Paul Samuelson wrote: "An 'incomes policy' is needed to supplement fiscal and monetary policy—in order to give the mixed economy a better long-run Phillips curve or natural rate....Benign neglect, governmental guideposts (voluntary or quasi-voluntary), direct wage–price controls, centralized collective bargaining, stop–go driving of the economy to cool it down, labor retraining programs to lower the natural level and range of structural unemployment—all these need study to retain the humane qualities of the modern order while achieving efficiency and stability" (Samuelson 1980, 781–82). The idea that central banks should be held accountable for inflation and could control it through control of the money stock was still being debated.

But, as the Friedmans noted in the closing chapter of *Free to Choose*, the tide was turning. Margaret Thatcher was elected prime minister of the United Kingdom in a landslide election in May 1979. With a solid majority in the House of Commons, Thatcher began a program of rolling back state involvement in every aspect of economic life in the UK. Large parts of Britain's industrial base that had been nationalized under earlier Labour and Conservative governments were privatized, starting with British Telecom in 1984 and followed by British Gas in 1986, British Airways and Rolls Royce in 1987, and a slew of others through British Rail and British Energy in the mid-1990s. The deregulation of the U.S. economy that had begun with airlines in 1977 accelerated under Ronald Reagan, who was elected president in 1980. Reforms had begun in China in 1978 under Deng Xiaoping's leadership, starting with a revival of private farming. Reforms were beginning in Latin America, with Chile leading the way in a number of areas.

These reforms were not exclusively the province of conservative parties. In New Zealand the Labour government of Roger Douglas embarked on a series of reforms that became a model for many other countries. New Zealand pioneered the idea of inflation targeting as a strategy for monetary policy that would focus central banks' policy deliberations and hold them accountable for inflation outcomes. This prescription for monetary policy has become increasingly popular in recent years and addresses many (though not all) of the Friedmans' concerns about discretionary monetary policy.

While inflation was close to a postwar peak in 1980, efforts were under way to bring it under control. In August 1979, Paul Volcker was appointed chairman of the Federal Reserve System, and the Fed embarked on a campaign to bring inflation down. By the time Volcker left office in 1987, inflation had fallen from 13.5 percent to 3.6 percent. Under Alan Greenspan's leadership, the Fed kept inflation under control and indeed lowered it further, to the point that by the beginning of the twenty-first century most commentators had stopped worrying about inflation and instead started worrying about deflation.

THE WORLD IN 2004

In the two-plus decades since *Free to Choose* was published, the world has changed dramatically, and in most ways for the better. There is less government involvement in most aspects of economic life than there was twenty-five years ago, inflation is lower, global trade is freer, and by most measures more people enjoy more economic freedom than at any time in the recent past. Living standards have risen for most of the world's population. The great experiment of the twentieth century has ended: The liberal, free-market, democratic model won. The Soviet Union has ceased to exist, and communism is no longer viewed as a viable alternative to free market capitalism. Russia is in the process of becoming a free market democracy. China, while still ruled by the Communist Party, has opened further to the world and has grown at rates that will make it the world's largest economy within a couple of decades. In May 2004 the European Union expanded from 15 to 25 members, incorporating many of the former Eastern European vassal states of the Soviet empire and in the process becoming one of the largest free trade blocs in the world.

Living standards around the globe are dramatically higher than they were twenty years ago, helped by the rolling back of the state in many countries and the lifting of restrictions on domestic and international trade (globalization). The fraction of the world's population living on less than $1 a day has fallen from 31.5 percent to 23.7 percent. While the number of people living in poverty remains large, there is greater acceptance that the surest way out of poverty is the protection of property rights, rule of law, and freedom to transact.

But if there have been great gains around the world, here in the United States progress has been slow in an area desperately needing reform. As **Eric Hanushek** observes in his paper, it has proven easier to defeat the forces of communism than to overcome the education establishment's resistance to meaningful reform of the public school system. The idea that school choice is essential to improving school quality is central to both *Capitalism and Freedom* and *Free to Choose*. School choice is so important to the Friedmans that their foundation is dedicated to promoting school choice and nothing else.[7] In *Free to Choose* the Friedmans wrote, "We believe that vouchers or their equivalent will be introduced in some form or other soon. We are more optimistic in this area than in welfare because education touches so many of us so deeply. We are willing to make far greater efforts to improve the schooling of our children than to eliminate waste and inequity in the distribution of relief. Discontent with schooling has been rising. So far as we can see, greater parental choice is the only alternative that is available to reduce that discontent. Vouchers keep being rejected and keep emerging with more and more support" (*Free to Choose*, 175). In the intervening years, of course, the United States has undertaken a far more radical reform of the welfare laws than has been attempted in education. As Hanushek shows, the performance of U.S. public schools has at best been stagnant, despite a massive increase in the resources available to them. U.S. students continue to perform poorly against students in other countries on standardized tests. This must surely be a source of continued concern in our increasingly integrated global economy.

But there has been some progress. The first major voucher program in the United States started in Milwaukee in 1990, and by the end of the decade some

10,000 students were participating in the program. The only other major school voucher program began in Cleveland in 1996–97. In 2002 the Supreme Court upheld the constitutionality of the Cleveland program, but there is still considerable opposition to expanding such programs to other school districts. An alternative means for promoting school choice is the charter school movement. Since the first charter school was authorized in Minnesota in 1991, some 2,700 charter schools have been opened in 36 states. And the No Child Left Behind Act of 2001 will contribute to greater school accountability and transparency and further the cause of reform, not least by making more and better information available to parents.

None of the constitutional amendments offered by the Friedmans in the concluding chapter of *Free to Choose* have been adopted in the United States. In his contribution, **Allan Meltzer** counts some twenty-five specific policy proposals in *Capitalism and Freedom* and *Free to Choose*, some of which have been adopted and many of which have not. The unequivocal successes are the ending of the draft, the floating of the dollar, and the abolition of interest rate ceilings on bank deposits. There have also been partial successes in the lowering of tariff barriers around the world, deregulation of various industries, and the introduction of an element of competition in education. The Earned Income Tax Credit can be viewed as a step toward the negative income tax the Friedmans proposed as an alternative to the various welfare programs. Meltzer argues that free market solutions to various problems are more likely to be adopted if they have been articulated in advance of any crisis that might precipitate a major reform. This allows proponents of the policies to respond to criticisms and allows officials to acquire familiarity with proposals to the point of believing that they might work. And therein lies one of the most enduring contributions of *Capitalism and Freedom* and *Free to Choose*.

In the wake of 9/11, defense and security spending has increased significantly in the United States. Airport security, once the province of private firms, is now in the hands of a federal agency, the Transportation Security Administration. Just how big should government be? The Friedmans have always accepted that there is some limited role for government. In *Free to Choose*, they quote from Adam Smith's *Wealth of Nations* to define the appropriate tasks of government as being

> first, the duty of protecting the society from the violence and invasion of other independent societies; secondly, the duty of protecting, as far as possible, every member of the society from the injustice or oppression of every other member of it, or the duty of establishing an exact administration of justice; and, thirdly, the duty of erecting and maintaining certain public works and certain public institutions, which it can never be for the interest of any individual, or small number of individuals, to erect and maintain; because

the profit could never repay the expense to any individual or small number of individuals, though it may frequently do much more than repay it to a great society. (*Free to Choose*, 28–29, quote from *Wealth of Nations*)

Indeed, as Raghurum Rajan emphasizes, the absence of government can be just as anticompetitive and detrimental to free markets as too much government. Governments today perform a much wider array of functions than those listed by Adam Smith. Government is intimately involved in the education system, in the provision of health care, and in the provision of income security through unemployment and social security programs. A key argument in *Free to Choose* was that government had grown well beyond the size necessary for the protection of liberties and needed to be scaled back. **William Niskanen** takes up the issue of the appropriate size of government in his paper. Niskanen's primary focus is on the economic burden of taxation, but as an aside he calculates that the optimal size of government relative to GDP in the United States is about 10 percent. At present, government expenditures account for more than 30 percent of GDP.

I have already alluded to the widespread belief in the 1930s that capitalism had failed as justification for greater government involvement in the economy. The response in the United States came in the form of the New Deal, which included the creation of the Social Security (Old Age and Survivors Insurance) program. This program is now the largest single item in the federal budget and accounts for more than a fifth of all federal spending. The changing demographics of the United States (falling birthrate and rising life expectancy) have made the system unsustainable in its current form. For a long time the issue was regarded as the third rail of U.S. politics, but there are signs that more politicians are willing to address the issue of the Social Security system's long-term solvency. In his paper, **Thomas Saving** and his co-authors document the size of the funding problem and analyze the costs and benefits of a transition from the current system to the Friedmans' preferred system of private accounts. Transitioning from the current publicly funded system to a privately funded one would make the country as a whole better off by enhancing the nation's capital stock. But such a transition will come at a cost in the form of lower consumption during the transition period.

By far the most dramatic development internationally since the publication of *Free to Choose* has been the collapse of almost all communist regimes in place in 1980. In his paper, **Peter Boettke** discusses the importance of the Friedmans' ideas in the reform process in the former communist societies. Many of the leading reformers had studied Friedman's work. The mass privatizations that took place in many of the former communist countries were inspired by Friedman's ideas. President Vaclav Klaus of the Czech Republic has acknowledged the importance of Friedman's ideas and intellectual courage to the reformers in Eastern Europe and has credited him with providing them "a clear vision where to go and a pragmatic strategy how to get there."

The expansion of economic freedom in China over the past quarter century is the subject of **Gregory Chow**'s paper. Chow documents the growth of economic freedom in China since the reform process started in 1978 and argues that this has contributed to an increase in political freedom as well. Government is still present in many areas of economic life, but its role is much diminished. Social insurance that was previously provided through guaranteed jobs in communes or state enterprises or health care through the same has been replaced by explicit programs providing unemployment, health, and old age insurance. Chow claims that there is probably a greater degree of freedom of choice in education in China than there is in the United States. He cites figures showing that some 40 percent of all spending on education in China comes from private sources versus an average of 12 percent for all Organization for Economic Cooperation and Development countries. Chow argues that "there appears to be no serious infringement of economic freedom in China, with the exception of the one-child policy," although according to the most recent report of the Economic Freedom Network (Gwartney, Lawson, and Emerick 2003), China ranked 100th out of 123 countries considered, with a score of 5.5 out of a possible 10. Hong Kong has consistently ranked at or near the top of all rankings of economic freedom. Chow comments on inflation's role in the Nationalist government's downfall in 1949 and in the unrest that culminated in the Tiananmen Square protests in 1989. One aspect of Friedman's thinking influenced policymakers in China even before the 1978 reforms: Apparently even the Marxian economics textbooks used in China's universities contained the quantity equation.

Green economics was just on the horizon when *Free to Choose* was published. In their autobiography, the Friedmans write that they had contemplated including "Pollutions" as one of the topics to be addressed in the TV series on which the book is based. Chapter 7 of *Free to Choose*, titled "Who Protects the Consumer?", has a brief discussion of environmental issues, and the Friedmans observe that the environmental movement has been behind a lot of the growth in government intervention in the economy. In the years since, the environmental movement has gathered strength, and environmental issues usually top the list of concerns of antiglobalization protesters. In their papers, **Terry Anderson** and **Richard Stroup** address environmental issues from a free market perspective. Anderson points out that countries with greater economic freedom and rule of law tend to have higher environmental standards (as measured by water pollution and so on) than countries in which the rule of law is weak. Indeed, the great level of wealth that economic freedom makes possible is itself a contributor to better environmental standards. There appears to be a Kuznets curve relationship between environmental quality and per capita GDP: At low levels of output, environmental quality deteriorates as countries trade off environmental quality for faster growth, but as output rises, societies demand and can afford cleaner environments. Contrary to the beliefs of many environmentalists, it is not necessary to have the government involved to ensure a better environment. Well-defined property rights and rule of law are all that is necessary to protect the environment from "tragedy of the commons" outcomes. Stroup explains why government regulation to achieve environmental objectives typically does worse than private property rights and free markets. To begin with, regulators typically will not have access to the information generated by and available to participants in free markets. And regulators will have little incentive to obtain that information by other means. Second, Coasian bargaining will generally ensure that a property right will flow to the highest-value user, but such exchange is often prohibited in a regulatory setting. Third, decisions made in the public sector are public goods, and there is limited accountability. Finally, competition leads to better quality goods, whether in education, as we have already seen, or in the environment. Public-sector entities that are not subject to competitive pressures will be less inclined to innovate than private-sector entities producing the same goods.

IS THE TIDE TURNING AGAIN?

The last chapter of *Free to Choose* is titled "The Tide Is Turning," and there is no doubt that the last quarter of the twentieth century saw a significant increase in economic freedom around the world. But many are now wondering if the tide is turning yet again, this time toward less economic freedom. Above, we noted that New Zealand was one of the first countries to fully embrace the idea of the need to roll back government intervention in the economy. Despite widespread pro-market reforms, since 1984 New Zealand has experienced one of the slowest growth rates of per capita GDP in the developed world. The sense that the market reforms had failed to deliver contributed to the election of a government that in 1999 started to roll back some of the previous decade's reforms. Trade unions have been given more power in wage negotiations, and the top income tax rate has been increased.[8]

In the decade following the publication of *Free to Choose*, U.S. government purchases as a percentage of GDP hovered in the 20–21 percent range. However, following the collapse of the Berlin Wall in 1989, government purchases as a fraction of GDP began a steady decline and bottomed out at 17.4 percent in 1998. Since then, the fraction of aggregate output absorbed by the government has increased each year, rising to 18.4 percent by 2003. Government expenditures as a fraction of GDP (which include transfer payments in addition to spending on consumption and investment) displayed stronger growth over the entire postwar period, peaking at 32.4 percent in 1992 before beginning a steady downward trend for much of the 1990s. However, this trend was also reversed in 2000. These are but two very crude measures of the government's overall impact on the U.S. economy. Other measures tell a similar story: The number of pages in the *Federal Register*, which Friedman has often used to gauge the extent of government involvement in the U.S. economy, reached an all-time record of 75,606 pages in 2002, an increase of about 9 percent over 2001.

Signs indicate that the enthusiasm of some countries for market-friendly reforms is waning, especially in Latin America. After a decade of significant rollbacks of the state in many Latin American countries, recent years have seen a backlash against so-called neo-liberalism. Brazil, Latin America's largest democracy, elected an avowed populist in 2002, as did the electorate of Argentina. The collapse of the convertibility plan in Argentina is seen as discrediting many of the reforms pioneered by former President Carlos Menem and his erstwhile finance minister Domingo Cavallo. Many in Latin America now talk of "reform fatigue." In some cases this is because the reforms were in name only, carried out less to spread private property as widely as possible than to enrich established interests.

The success of some countries that have grown rapidly as they have become more integrated into the global trading system has begun to provoke a backlash in the more developed countries. In recent years, there has been a growing outcry against outsourcing of jobs from the United States and Europe to countries such as China and India. Large sectors of the economies of the advanced industrial countries were previously thought immune to foreign competition because their products were nontradable across national borders. As more countries open to trade and as technology makes many services internationally tradable, workers are finding that employment in these sectors is less secure. Increasingly, white-collar workers are joining blue-collar workers in questioning the benefits of free trade.

In short, we cannot take for granted the progress toward greater economic freedom that we have seen over the past two decades. There is nothing inevitable about such progress, and history teaches us that the process can be reversed, with dire consequences. The liberal economic order that existed in most of the world before World War I was destroyed in the turmoil of the interwar years, and it took decades for markets to be reopened.[9] While goods, services, and capital can now flow between countries with the same ease as in the pre-World War I period, the same is not true of people. There are still large barriers to international migration (except within the European Union), due in part to the postwar creation of welfare states in most of the advanced economies. The voices of the critics of economic liberalism grow louder every day, and in his paper, Raghurum Rajan notes the need to "engage dissident economists and demagogic activists in fruitful dialogue, instead of letting them dominate the public arena." Unfortunately, few contemporary academic economists are willing to take up this challenge.

CONCLUSION

In his remarks to the conference, **Alan Greenspan** repeated the famous quotation from John Maynard Keynes' *General Theory* on the power of ideas: "The ideas of economists and political philosophers, both when they are right and when they are wrong, are more powerful than is commonly understood. Indeed, the world is ruled by little else. Practical men, who believe themselves to be quite exempt from any intellectual influences, are usually the slaves of some defunct economist. Madmen in authority, who hear voices in the air, are distilling their frenzy from some academic scribbler of a few years back." The influence of Milton and Rose Friedman on the course of the late twentieth century is testimony to the power of ideas and the ability of two individuals to make a difference. Friedman was the dominant figure in the University of Chicago's economics department for thirty years, and to many he is the figure most closely identified with what came to be known as the Chicago school.

But his influence extends well beyond economics. Former Secretary of State George Shultz has described Milton Friedman as the individual who has had the most impact on the modern world. Friedman's enormous influence on public policy stemmed not just from the quality of the scientific work for which he was awarded a Nobel Prize in 1976, but also from the fact that he was willing to go out and argue in public forums for the benefits of free market capitalism at a time when it was distinctly unfashionable to do so. Milton and Rose Friedman describe themselves in their autobiography as two lucky people. They were on the right side of the great debate of the twentieth century, and they had the good fortune to see their arguments vindicated by the course of experience.

NOTES

[1] *Capitalism and Freedom* was reviewed by John Hicks in *Economica*, Paul Baran in *Journal of Political Economy*, and Abba Lerner in *American Economic Review*.

[2] In 1972 a group of Friedman's former students organized a conference at the University of Virginia to re-examine the ideas in *Capitalism and Freedom* to celebrate Friedman's sixtieth birthday. The conference proceedings were subsequently published as *Capitalism and Freedom: Problems and Prospects. Proceedings of a Conference in Honor of Milton Friedman*, ed. Richard T. Selden, University Press of Virginia, 1975. Two of the participants in that earlier conference also participated in the Federal Reserve Bank of Dallas conference thirty years later (William Niskanen and Gary Becker).

[3] Becker's speech is not included in this volume.

[4] Friedman (1986) provides an interesting perspective on the resource costs of fiat monetary standards.

[5] See also Friedman's interview with the *Financial Times*, June 5, 2003, and his opinion piece in the *Wall Street Journal*, August 19, 2003.

[6] It is interesting to reread some of the Friedmans' commentary on the run-up to the Great Depression in the light of recent economic history: "The high tide of the [Federal Reserve] System was undoubtedly the rest of the twenties. During those few years it did serve as an effective balance wheel, increasing the rate of monetary growth when the economy showed signs of faltering, and reducing the rate of monetary growth when the economy started expanding more rapidly. It did not prevent fluctuations in the economy, but it did contribute to keeping them mild. Moreover, it was sufficiently evenhanded so that it avoided inflation. The result of the stable monetary and economic climate was rapid economic growth. It was widely trumpeted that a new era had arrived, that the business cycle was dead, dispatched by a vigilant Federal Reserve System" (*Free to Choose*, 78).

[7] For information on the Milton and Rose D. Friedman Foundation, see www.friedmanfoundation.org.

[8] See "Can the Kiwi Economy Fly?" *The Economist*, November 30, 2000.

[9] A good reference is James (2001).

REFERENCES

Friedman, Milton. 1986. "The Resource Cost of Irredeemable Paper Money." *Journal of Political Economy* 94: 642–47.

———, with the assistance of Rose D. Friedman. 1962. *Capitalism and Freedom*. Chicago: University of Chicago Press.

Friedman, Milton, and Rose D. Friedman. 1980. *Free to Choose: A Personal Statement*. New York: Harcourt Brace Jovanovich.

———. 1998. *Two Lucky People*. Chicago: University of Chicago Press.

Friedman, Milton, and Anna J. Schwartz. 1963. *A Monetary History of the United States, 1867–1960*. Princeton: Princeton University Press.

Gwartney, James, and Robert Lawson, with Neil Emerick. 2003. *Economic Freedom of the World: 2003 Annual Report*. Vancouver, BC: The Fraser Institute.

James, Harold. 2001. *The End of Globalization: Lessons from the Great Depression*. Cambridge, MA: Harvard University Press.

Keynes, John Maynard. 1936. *The General Theory of Employment, Interest and Money*. New York: Harcourt, Brace and Co.

Samuelson, Paul A. 1980. *Economics*, 11th ed. New York: McGraw-Hill.

Session 1

The Toughest Battleground: Schools
Eric A. Hanushek

The Theory and Practice of School Choice
Paul E. Peterson

The Toughest Battleground: Schools

Eric A. Hanushek

Over four decades ago, Milton Friedman published *Capitalism and Freedom* (Friedman 1962). This insightful little book traveled across a broad range of important topics collected around the theme of how government can best operate within a free society. The message was expanded two decades later in *Free to Choose* (Friedman and Friedman 1980). At the time, the battle of the ideas introduced by these books was being waged by nations, nations that were willing to contemplate war over how societies should be organized. As we look back on how the world has changed since then, I wonder if anybody guessed that changing the schools would be the most difficult subject taken on. It is useful to look at what progress has been made, what evidence exists on the topic, and what the future might hold in the area of education. The simple question is: Why are the schools tougher to crack than the walls of the Communist bloc?

ARGUMENTS ABOUT SCHOOLS

Perhaps the key insight in *Capitalism and Freedom* was that government concern about schools and the schooling of its population could be separated from the issue of who actually runs the schools. While government may want to finance schools for a variety of reasons—externalities, economies of scale, income distribution, or what have you—they do not have to do the actual production. Indeed, there are reasons—obvious now, but perhaps not as obvious forty years ago—why government monopoly in schools may be undesirable. These themes are amplified in *Free to Choose*.

In my economics of education course, Friedman's chapter 6, "The Role of Government in Education," occupies an early lesson. And perhaps no other section of the course incites such eye-opening thoughts and raw emotions as discussion of vouchers and private schools perhaps replacing some public schools. This paper reviews what has happened in schools since 1962, chronicling the somewhat divergent paths of schooling outcomes and the intellectual debate on choice.

Before doing this, however, it is important to underscore what may be an equally important chapter of the book for the activities of schools, the chapter on occupational licensure. Today, while battles continue around ideas of choice in schools, equally strong and, in my view, potentially equally damaging battles surround the appropriate standards for credentialing and licensure of teachers.

SOME FACTS ABOUT U.S. SCHOOLS SINCE 1960

The backdrop for today's discussion is what has happened to U.S. public schools over the time since *Capitalism and Freedom* and *Free to Choose* entered into the intellectual fabric of the country.

Start with the resources and support for public schools. Table 1 displays the pattern of resources supplied to public schools in the United States between 1960 and 2000. Several things are obvious from this table. First, the United States has been running a class size "experiment" for forty years. Between 1960 and 2000, the pupil–teacher ratio fell by more than a third. Second, there has also been an expansion in the conventional measures of teacher quality—graduate education and experience. The percentage of teachers with a master's degree or more doubled over this period, with the typical teacher now having an advanced degree. Experience also reached new heights.

An obvious implication of these changes in real resources of schools is that spending on schools has risen dramatically. Teacher education and experience are prime determinants of teacher salaries, and the pupil–teacher ratio determines across how many students the salaries are spread. Thus, as the last line of the table indicates, real spending per pupil in schools was *240 percent higher* in 2000 than in 1960. That is, after adjusting for inflation, we had truly dramatic increases in our school spending—increases that appear to exceed public perceptions by a wide margin.

Table 1
Public School Resources in the United States, 1960–2000

	1960	1980	2000
Pupil–teacher ratio	25.8	18.7	16.0
Percentage of teachers with master's degree or more	23.5	49.6	56.2*
Median years teacher experience	11	12	15*
Real expenditure/ADA (2000–01 dollars)	$2,235	$5,124	$7,591

* Data for 1996.
SOURCE: U.S. Department of Education (2002).

The contrast of resource increases with what has happened to student performance is equally startling. Figure 1 displays the pattern of performance of U.S. 17-year-olds from the National Assessment of Educational Progress (NAEP). NAEP provides a consistent measure of performance over time for a random selection of students. The picture shows that mathematics and reading performance is up slightly over the period while science is down.[1] A simple summary of performance over this period is that it was flat. School resources more than tripled, but there was no discernible effect on performance.

Of course, the overall trends could be misleading, particularly if there were significant changes in the student population or in the institutional structure of schools. For example, it is frequently cited that families are less stable or that there are more difficult-to-educate immigrants in the schools. Indeed, until the decade of the 1990s, the proportion of children in poverty had been rising. Relatedly, the proportion of children in single-parent families has risen, although this leveled off in the last decade. Finally, in terms of factors adversely affecting achievement, the prevalence of families not speaking English at home has increased.

But, these adverse changes have coincided with other changes that would generally be viewed as favorable for children and learning. Parents are more educated, and families are smaller. Additionally, greater percentages of children age four and five are attending preschool programs.

Figure 1
Performance of 17-Year-Olds on the National Assessment of Educational Progress (NAEP), 1960–99

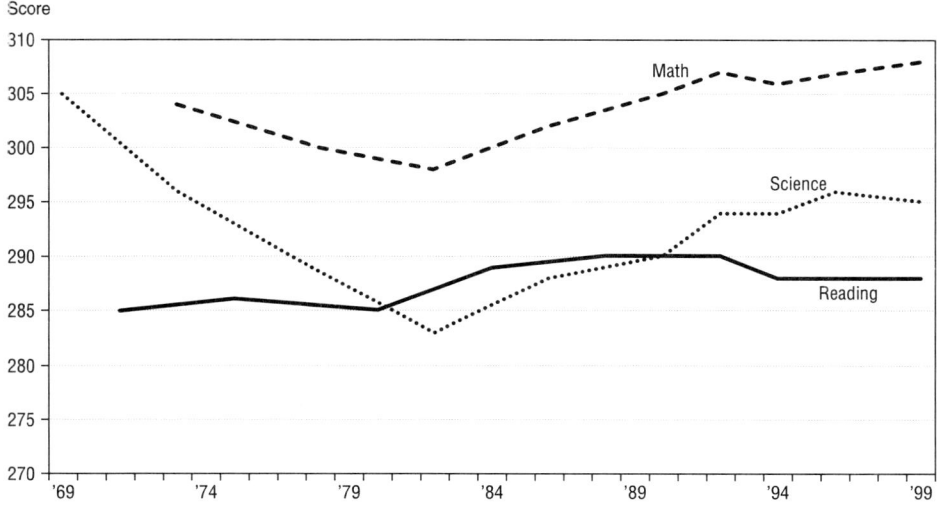

It is difficult to know precisely how these factors net out in their overall effect on students. The best estimates available, while surrounded by uncertainty, suggest that the net effect of these factors is, if anything, positive. Without taking a strong stand, it is sufficient to conclude that the evidence does not show an overwhelming decline in "student input quality."

In sum, the aggregate data do not suggest that the existing schooling system has been performing very well, even though resources have been provided at sharply increasing levels over the past decades.

There is, nonetheless, a different possible perspective. If U.S. performance has been high and has exceeded that of other nations, the fact that it is flat over time might not be such a concern. In that case, the main issue to be considered here would be the continual pressures to increase expenditure (with the implication that inefficiency in government provision of schooling has been increasing). Unfortunately, that interpretation does not hold up. Table 2 shows the U.S. ranking on international math and science examinations given in 1995. The results of the Third International Mathematics and Science Study (TIMSS) show that particularly by the twelfth grade U.S. students are simply not competitive with those from other countries—ranking 19 and 16 out of the participating 21 countries in mathematics and science, respectively.

It is increasingly difficult to resist the conclusion that U.S. public schools are not performing particularly well. Perhaps in the past it could have been argued that with a little more time, with a few more resources, with adoption of today's good ideas, things will get much better. At some point, though, we have to face reality.

But let us look at the other side of the story. Have families abandoned the public schools, seeking out better opportunities elsewhere? First, in terms of private-school enrollment, the answer is essentially no. The percentage of students in private schools has fallen since 1959 and has remained stuck at roughly 11 percent of the K–12 student population. The largest change has been the composition of the religious schools. Catholic schools have gone through a long decline, being replaced by other religious schools. Private, nonsectarian schooling has remained roughly constant.

One aspect of the schools, however, points in a slightly different direction. We have seen the rise in homeschooling—i.e., complete withdrawal and a

Table 2
U.S. Rank on the Third International Mathematics and Science Study (TIMSS), 1995

	Fourth Grade	**Eighth Grade**	**Twelfth Grade**
Mathematics	12 out of 26	28 out of 41	19 out of 21
Science	3 out of 26	17 out of 41	16 out of 21

return to the schooling that preceded the public schooling movement in the United States. The data on this are sketchy. It is difficult to track even the numbers of students in homeschooling, let alone find out any information about the results of this. Some estimates suggest that as many as 2 percent of students of the relevant K–12 age population are being homeschooled, but it might be 1.5 percent or 2.5 percent (Henke et al. 2000; Bielick, Chandler, and Broughman 2001).

Capitalism and Freedom suggests that one reason advanced for the current governmental role in schools is that of "technical monopoly." It may not be possible to elicit sufficient supply of private schools if there is low population density coupled with some economies of scale, at least at very small school sizes. And yet, a significant number of parents are choosing the very expensive option of schooling their own children because they find that the public schools are not meeting their demands.[2]

One other trend has begun to intrude on schooling in the United States. A wide range of analyses, including the influential *A Nation at Risk* in 1983, have suggested that U.S. schools face serious problems. Partly related to a continued desire to "reform" the schools, a number of policy discussions focus on the importance of high-quality teachers. This secondary "reform" effort (in addition to simply supplying more resources) has been tightening up on the credentialing of teachers. New and more restrictive requirements for teacher credentials have been introduced in a variety of states and are contemplated in even more. The only thing absent has been any demonstration that these new requirements are at all related to the ability of teachers to improve student achievement (Hanushek and Rivkin 2004).

THE FACETS OF RESISTANCE

Does the continued draw of the public schools represent a refutation of Friedman arguments that there is pent-up demand by households for schools that look different from the public schools? Hardly.

First, throughout this entire period, with small exceptions discussed below, parents have a choice between free public schools and costly private schools. Moreover, because the costs of schools are spread across the whole population, the resources available in the public schools generally exceed the tax expenditures of parents. Privately matching these expenditures represents substantial expenditures by parents.

Second, parents have been systematically led to believe that their schools are doing quite well. Regularly, the typical parent rates his or her own school as a B-plus, even if the very same random sample believes that the other public schools rate a C-minus (Rose and Gallup 2001). Perhaps because the typical parent learned math in the public schools, few worried about this inconsistency, at least before Garrison Keillor publicized Lake Wobegon. Moe (2001) also finds that a small proportion of all parents think their schools are in need of serious change, and parent satisfaction with schools rises with family income.

There is also an aggregate story. A variety of writers who do not want to see any fundamental change in the structure and operation of schools simply take the position that all is well. Consider Alfie Kohn, a prominent critic of academic standards and testing, who wrote in 2000:

> As proof of the inadequacy of U.S. schools, many writers and public officials pointed to the sputtering condition of the U.S. economy. As far as I know, none of them subsequently apologized for offering a mistaken and unfair attack on our educational system once the economy recovered, nor did anyone credit teachers for the turnaround.[3]

Another prominent defender of the school system, Gerald Bracey, took the argument one step further. Noting that a variety of people from before and after *A Nation at Risk* had argued for improving schools in order to maintain U.S. economic strength, he wrote:

> None of these fine gentlemen provided any data on the relationship between the economy's health and the performance of schools. Our long economic boom suggests there isn't one—or that our schools are better than the critics claim.[4]

Of course, what these authors have ignored is that the idea behind improving the quality of schools is a long-run issue. Improving the knowledge of today's high school seniors will obviously not translate into lower unemployment today. In fact, it will not be discernible for some time to come. Moreover, a range of other features of the U.S. economy enter into economic growth and the productivity of the nation's labor force. Indeed, these are in part the other elements of *Capitalism and Freedom*: the comparatively favorable U.S. regulatory environment, the limited size of government, and the fewer intrusions in labor and product markets. (Here, however, I am sure that Professor Friedman will rightfully say, "Maybe, but these things could be better.")

But, this discussion of resistance to change cannot be complete without noting a truly significant change in schooling. In the early 1960s, the idea of teachers overwhelmingly joining in a traditional craft union was not really on the horizon. The early debates about unionization—as opposed to simple professional organizations—did not clearly point to the current heavily unionized teaching force.

One aspect of this unionization is the concentration of immense political power. There are currently three million teachers, a significant voting bloc with very specific ideas on the form that any reform of schools should take. The teachers unions control significant political funds (coming directly from union dues), and they use these funds to further their agenda at the local, state, and federal level.

Picture the District of Columbia. This is an odd school district, because it comes directly under the jurisdiction of the U.S. Congress. In 2000, D.C. spent $10,874 per pupil, dramatically above the average spending for the country, which is less than $7,000. Yet, according to the National Assessment of Educational Progress, performance of D.C. students rated last in the nation. Moreover, performance of just black students in D.C. compared with black students in other states also ranked at the very bottom. Imagine then that some members of Congress, including the representative for the District of Columbia, are trying to obstruct the introduction of a voucher at an amount noticeably less than the current average spending in the District. The argument: We should not do anything to harm the public schools of the District.

One just has to witness the amount of opposition spending by the unions to the voucher referenda in California, Michigan, and Colorado to have an appreciation for the self-interested politics. The very sophisticated media campaigns supported by the teachers unions convinced voters that the introduction of vouchers, no matter how constrained, would damage the public schools, would be expensive, and would not be in the interests of society.

EXPERIENCE WITH VOUCHERS

A few cracks have developed in the resistance to vouchers. These include the introduction of a limited voucher program in Milwaukee, the broader introduction in Cleveland along with the U.S. Supreme Court affirmation of such policies, and the introduction of a variety of private voucher programs.

These experiences have been discussed and analyzed in many different places (e.g., Rouse 1998, Howell and Peterson 2002) and are set out in Paul Peterson's commentary for this conference (Peterson 2004). While different authors and commentators have interpreted the data differently, my summary is fairly straightforward.

First, none of these are general tests of voucher programs. They rely (at least until recently) on schools in existence before the vouchers were introduced. Thus, they give little indication of any supply response that might be seen if there were more general vouchers that were assured of being around for some time into the future.

Second, in almost all situations the expenditures in the voucher schools are noticeably less than those in the competing public schools. This differential implies that these are not tests of Friedman vouchers, although they may give some partial information.

Third, parents tend to be happier with the voucher schools than with the corresponding public schools (Witte 1999, Howell and Peterson 2002, Peterson

2004). In other words, even given the restrictions discussed above, there is a group of parents that highly values the alternative schools.

Finally, achievement in voucher schools appears to be as high as or higher than that in comparison public schools. Allowing for possible differences in student bodies, those attending voucher schools score better on average—although this is not consistent across subgroup, outcome measures, or length of voucher operation.

Before evaluating these results, however, it is useful to expand the discussion to include other forms of choice.

EXPERIENCE WITH OTHER FORMS OF CHOICE

One of the significant changes in school choice since *Capitalism and Freedom* has been the introduction of different kinds of school choice. While vouchers are the purest form, and the one obviously preferred by the Friedmans, innovation in choice has occurred.

As previously mentioned, there has been a considerable surge in home-schooling. A significant number of parents have simply withdrawn their children from the regular public schools and taken personal responsibility for their education. Unfortunately, however, little is known about this in terms of movements in and out or of performance.

Citizen sentiment for expanded choice has generally increased over time, a fact not missed by opponents of more choice. Thus, one reaction to calls for vouchers and more choice has been the mantra of a number of people that they are for choice but it should be restricted to public school choice. This position has been particularly popular among politicians who want to protect the existing public schools from any competitive pressures while still seeming open to more fundamental reforms.

A particularly popular version of public school choice involves an open-enrollment plan. For example, students could apply to attend a different school in their district rather than the one to which they are originally assigned. Or in a more expansive version, no initial assignment is made, and students apply to an ordered set of district schools. A common version of this has been the use of magnet schools that offer a specialized focus such as college preparatory or the arts.[5]

It is fair to say that these public school choice plans do not even bear a pale resemblance to the ideas of choice included in voucher plans. First, the flows of students are heavily controlled. For example, the first caveat is always "if there is space at the school," but the desirable public schools virtually never have space. Second, large urban school systems where there is a natural range of options frequently face other restrictions, such as racial balance concerns that severely constrain the outcomes that are permitted.

Third, and most important, these plans seldom have much effect on incentives in the schools. The competitive model of vouchers envisions that schools that are unable to attract students will shut down. This threat provides an incentive to people in the schools to perform well or potentially lose their jobs. But in the cases of open-enrollment schools that are not fully regulated to ensure that all of the schools maintain enrollment, the people in undersubscribed schools generally still have employment rights and would simply move to another school with more students.

Some magnet schools do look to be very good and almost certainly meet the interests of the attendees. The long illustrious records of Brooklyn Tech, Stuyvesant, and Bronx Science in New York City stand out. But this is far different from the idea of introducing more competition in the provision of schools.

A variant of open-enrollment plans is permitting students in a city to attend any public school in a state. Conceptually this could offer some competitive incentives. If a district lost sufficient students through out-migration, they could be left with less funding and could be forced to reduce their workforce. Again, however, the reality is not much in the way of competition. The funding for such plans generally has the choice student carrying less than the full funding for the receiving district, meaning that any district accepting students is asking its residents to subsidize the education of students outside the district. Further, the "if there is space at the school" clause generally stops all but some token movement.

A different development looks closer to voucher schools—that of charter schools. There is no common model for charter schools because they are creatures of the separate states and operate in different ways according to state rules. The essential features are that they are public schools, but ones that are allowed to operate to varying degrees outside of the normal public schools. They are schools of choice, surviving through their ability to attract sufficient numbers of students. Their form differs widely, however, in the rules for their establishment, in the regulations that apply to them, in the financing that goes with the students, and in a host of other potentially important dimensions (see, for example, Finn, Manno, and Vanourek 2000). Some states, for example, layer a variety of requirements about teacher certification, curriculum, acceptance of special education students, and the like—advertised as "leveling the playing field"—to ensure that charter schools do not offer any true innovation and competition. Other states, however, remove a substantial amount of regulation and truly solicit innovation and competition (Center for Education Reform 2003).

Charter schools can offer true competition to the regular public schools because they can draw students away from poorly performing regular publics. Employment rights typically do not transfer between charters and regular publics, so there is potentially pressure on school personnel to attract students.

Moreover, we see that a substantial number of attempted charters do not succeed in the marketplace (Center for Education Reform 2002).

Currently, some forty states and the District of Columbia have authorized charter schools. The student population attending charters has grown to over 1 percent nationally during the last decade, but in some places the enrollment is truly much more significant. For example, in the 2001–02 school year, 9.2 percent of students in the District of Columbia, 6.7 percent in Arizona, 3.8 percent in Michigan, and 3.7 percent in California attended charter schools (Hoffman 2003).

What do we know about the performance of charter schools? Analysis has actually been very limited. To begin with, any school of choice—from the classic Catholic schools through vouchers and charters—necessarily has a self-selected population.[6] Thus, inferring the impact of the school, as distinct from the characteristics of the students that are attracted, is always difficult. Additionally, because charter schools are largely new, most are still going through a start-up phase. The results observed during this phase may not be indicative of what they will look like in the steady state.

With those caveats in mind, I can provide some preliminary estimates of the performance of charters in Texas. Texas has a significant number of charter schools (although the legislature has capped the total number). Because Texas has tested students for a decade, it is possible to trace the students who enter and leave charter schools.[7] The simplest design that deals with the selection problems is to compare the average learning growth for individual students when in the regular public schools with their own performance in the charters. In this way, charter students become their own control group.

Preliminary results of Hanushek, Kain, and Rivkin (2003) address the issue of charter school quality. Three things come out of this in terms of quality indicators. First, on average, charter schools perform very similarly to the regular public schools. But, second, start-up problems are real, and new charters do not perform as well as more-established charters. More-established charters (those over four years in age) on average outperform the regular public schools of Texas. Third, there is a significant distribution of performance across both regular publics and privates. The good are good, and the bad are truly bad.

Note that this judgment is also biased against charter schools to the extent that their objectives may not simply be developing the basic math and reading skills that are used in the analysis. If they have specialized purposes, no attention is given to those.

These results await, nonetheless, both the general maturation of more charter schools and the investigation of their performance in different settings.

One other aspect of charter schools deserves mention. Choice schools have potential advantages by allowing students to find schools that meet their own interests and needs. But another important aspect of competitive markets is enforcing a discipline on the other participants—in this case, the regular public schools. Is there any evidence that the regular public schools respond to the pressures of competition? Again, it is very early in the development of charters, but Hoxby (2003) introduces preliminary evidence that there are competitive improvements.

One final result of the analysis of charters is important. If we look at the behavior of parents, we find that they are significantly more likely to withdraw their children from a poorly performing charter as compared with a well-performing charter. This finding is particularly important because parents are not given information on their charter school's value added.[8] The behavior of parents shows, however, that they are good consumers and that they can use the performance data that are available to infer the school's quality. An early and continual criticism of the voucher idea is that parents are not good consumers, an assertion belied by the data.

It is useful to note that parents make similar judgments about the regular public schools, but they are much less likely to exit a regular public school given bad performance. The reason is obvious: It is generally much more costly to change public schools, given that a change of residence is frequently required. Further, this ability to exit a given public school is not shared equally by all parents. Middle- and upper-income parents have the resources to select among alternative districts, almost surely explaining their generally greater satisfaction with the public schools (Moe 2001). This differential ability is also a situation noted in *Free to Choose*.

PROSPECTS FOR THE FUTURE

Let us retrace the discussion. If we begin with the situation in 1962—when Milton Friedman laid out the reasons for and benefits of enhanced choice—no measures suggest that student outcomes have improved. On the other side of the ledger, real spending has more than tripled, leading us to the unmistakable conclusion that schools have become more inefficient. Yet there has been little take-up of Friedman's basic proposal to introduce vouchers.

Is the introduction of broader choice hopeless? I would argue not.

The discussion of school choice stimulated by *Capitalism and Freedom* has grown and penetrated the broad public. A majority of parents and citizens now believe that more choice is desirable (Moe 2001, 2002).

Coupled with that predisposition is the beginning of better accountability by schools. Recent federal legislation in the No Child Left Behind Act of 2001 requires all states to develop regular performance measures of student learning and to make these measures publicly available. As noted previously, the typical parent believes his or her child is attending an above-average school. What will happen when numbers of these parents learn they were wrong?

In my opinion, providing parents and policymakers with better information about the current shortcomings of their schools offers the possibility of breaking the schools loose from the stranglehold school establishment has on them. While I do not see that simple accountability will work without greater school choice, I also do not think we will easily arrive at much greater choice without strong accountability. This indeed is one of the clear messages from the Koret Task Force when it considered why there had been so little true improvement since *A Nation at Risk* (Peterson 2003).

This is also an important time. There are forces pulling in the exact opposite direction. The educational establishment has also argued that reform is needed, but their "reform" is very different. They argue for doing what we have been doing, just more of it. The movement to reduce class sizes, while slowed by the fiscal problems of states and localities, has not gone away. The teacher licensure forces are pushing for tightening up on credentials—requiring master's degrees of all, increasing the course requirements, deepening the ongoing professional development (Hanushek and Rivkin 2004). There is a struggle also to link tightened teacher credentials to the federal accountability requirements.[9] Substantial evidence suggests that improving the quality of teachers is key to any reform. There is no evidence to suggest that this will come from expanded certification and licensure.

It does not seem to matter that the portfolio of policy proposals emanating from the establishment looks much like those we have pursued over the past four decades. The only difference seems to be that those making these proposals disavow the mistakes of the past. They want to hear nothing of our schools' performance history. And they certainly do not acknowledge that the problems are deeper than being short on some standard dimensions.

Perhaps, however, we will still see the iron curtain that has surrounded school policy fall. The force—the same as in the economies of Eastern Europe—will be poor and inefficient performance.

NOTES

[1] Writing performance, not shown, was assessed between 1984 and 1996 and was significantly down over that period, although there are questions about the reliability of scoring the writing examinations. Longer time-series evidence on performance comes from the SAT test, which shows declines from the mid-1960s. This trend is difficult to interpret, however, because the SAT is a voluntary test, where participation rates have increased significantly over time. Nonetheless, analyses of these changes—particularly the earlier changes—suggest that the movement is a combination of decreased selectivity in test taking and real changes in skills and performance (Congressional Budget Office 1986, 1987).

[2] The reasons for choosing homeschooling are clearly complex. A survey of parents finds that half list "giving their children better education at home" as a reason, while 40 percent cite religious reasons (Bielick, Chandler, and Broughman 2001).

[3] Kohn (2000).

[4] Bracey (2002).

[5] Forms of open-enrollment plans were the response of a number of Southern districts to the desegregation orders flowing from *Brown v. Board of Education*. In general, simple open-enrollment plans were not found to satisfy the court requirements for desegregation of districts, but magnet schools (with racial balance restrictions) became a reasonably common policy approach (Armor 1995). In 2001–02, 3 percent of all students attended a magnet school (Hoffman 2003).

[6] For a more complete discussion of the analytical problem along with the evidence on Catholic schools, see Hanushek (2002).

[7] The ability to track students over time is the result of the Texas Schools Project at the University of Dallas. That project has linked students over time and matched them with their schools.

[8] In the previous analysis, the growth in student test scores was compared with that in regular public schools to obtain an estimate of each school's value added. Parents, however, can only observe an absolute score of student performance that is unadjusted for any selectivity of the school.

[9] No Child Left Behind requires that all students have a highly qualified teacher, a requirement that a number of groups are trying to equate to having existing or expanded credentials.

REFERENCES

Armor, David J. 1995. *Forced Justice: School Desegregation and the Law*. New York: Oxford University Press.

Bielick, Stacey, Kathryn Chandler, and Stephen P. Broughman. 2001. *Homeschooling in the United States: 1999*. Washington, DC: National Center for Education Statistics (July).

Bracey, Gerald W. 2002. "Why Do We Scapegoat the Schools?" *Washington Post*, May 5, 2002, B01.

Center for Education Reform. 2002. *Charter School Closures: The Opportunity for Accountability*. Washington, DC: Center for Education Reform.

———. 2003. *Charter School Laws Across the States: Ranking Scorecard and Legislative Profiles*. Washington, DC: Center for Education Reform (January).

Congressional Budget Office. 1986. *Trends in Educational Achievement*. Washington, DC: Congressional Budget Office.

———. 1987. *Educational Achievement: Explanations and Implications of Recent Trends*. Washington, DC: Congressional Budget Office.

Finn, Chester E., Jr., Bruno V. Manno, and Gregg Vanourek. 2000. *Charter Schools in Action*. Princeton, NJ: Princeton University Press.

Friedman, Milton. 1962. *Capitalism and Freedom*. Chicago: University of Chicago Press.

Friedman, Milton, and Rose Friedman. 1980. *Free to Choose: A Personal Statement*. San Diego: Harcourt, Inc.

Hanushek, Eric A. 2002. "Publicly Provided Education." In *Handbook of Public Economics*, ed. Alan J. Auerbach and Martin Feldstein, 2045–2141. Amsterdam: Elsevier.

Hanushek, Eric A., John F. Kain, and Steve G. Rivkin. 2003. "The Impact of Charter Schools on Academic Achievement." Paper presented at the annual meeting of the American Economic Association (January).

Hanushek, Eric A., and Steve G. Rivkin. 2004. "How to Improve the Supply of High Quality Teachers." In *Brookings Papers on Education Policy 2004*, ed. Diane Ravitch, 7–25. Washington, DC: Brookings Institution Press.

Henke, Robin R., Phillip Kaufman, Stephen P. Broughman, and Kathryn Chandler. 2000. *Issues Related to Estimating the Homeschooled Population in the United States with National Household Survey Data*. Washington, DC: National Center for Education Statistics (September).

Hoffman, Lee McGraw. 2003. *Overview of Public Elementary and Secondary Schools: School Year 2001–02*. Washington, DC: National Center for Education Statistics (May).

Howell, William G., and Paul E. Peterson. 2002. *The Education Gap: Vouchers and Urban Schools*. Washington, DC: Brookings Institution.

Hoxby, Caroline Minter. 2003. "School Choice and School Productivity (or Could School Choice Be a Tide That Lifts All Boats?)." In *The Economics of School Choice*, ed. Caroline Minter Hoxby. Chicago: University of Chicago Press.

Kohn, Alfie. 2000. *The Case Against Standardized Testing: Raising the Scores, Ruining the Schools*. Portsmouth, NH: Heinemann.

Moe, Terry M. 2001. *Schools, Vouchers and the American Public*. Washington, DC: Brookings Institution.

———. 2002. "Cooking the Questions." *Education Next* 2, no. 1 (Spring): 71–77.

Peterson, Paul E., ed. 2003. *Our Schools and Our Future: Are We Still at Risk?* Stanford, CA: Hoover Press.

———. 2004. "The Theory and Practice of School Choice." In *The Legacy of Milton and Rose Friedman's "Free to Choose": Economic Liberalism at the Turn of the Twenty-First Century*, ed. Mark A. Wynne, Harvey Rosenblum, and Robert L. Formaini, 37–54. Proceedings of a conference sponsored by the Federal Reserve Bank of Dallas, October 23–24, 2003.

Rose, Lowell C., and Alec M. Gallup. 2001. "The 33rd Annual Phi Delta Kappa/Gallup Poll of the Public's Attitudes Toward the Public Schools." *Phi Delta Kappan* (September): 41–58.

Rouse, Cecilia Elena. 1998. "Private School Vouchers and Student Achievement: An Evaluation of the Milwaukee Parental Choice Program." *Quarterly Journal of Economics* 113, no. 2 (May): 553–602.

U.S. Department of Education. 2002. *Digest of Education Statistics, 2001.* Washington, DC: National Center for Education Statistics.

Witte, John F., Jr. 1999. *The Market Approach to Education.* Princeton, NJ: Princeton University Press.

The Theory and Practice of School Choice

Paul E. Peterson

Economists prove in theory what works in practice. So it is said.

This paper demonstrates quite the opposite: It shows that school vouchers work in practice, just as Rose and Milton Friedman proved in theory.

Simply defined, a voucher is a coupon for the purchase of a particular good or service. Unlike a $10 bill, it cannot be used for any purpose whatsoever. Its use is limited to the terms designated by the voucher. But like a $10 bill, vouchers typically offer recipients a choice. For this reason, distant relatives find coupons popular birthday presents for those whose tastes are unknown.

It is not only in the retail market that vouchers or coupons are used. Food stamps, housing allowances for the poor, and federal grants for needy students are all voucherlike programs that fund services while giving recipients a range of choice. It is the special contribution of the Friedmans that they have shown, theoretically, how vouchers can also enhance school choice and school productivity. By giving parents a school voucher, the government ensures that the money will be used for an investment in human capital. But instead of requiring attendance at a government-operated neighborhood school, no matter how deficient, the family is given a choice among public and private schools in its community. Schools can then compete for customer support. If educational services do not differ significantly from other goods and services, then this market-based approach to educational provision should yield efficiency gains.

PUTTING THEORY TO A PRACTICAL TEST

Until the 1990s, there was little opportunity to rigorously test this application of classical economic theory to the provision of educational services. But in the past dozen years, small school voucher experiments have been initiated in a variety of places, providing a chance to see if educational practice does in fact conform to classical theory, as explicated and applied to school vouchers by the Friedmans.[1]

Publicly funded voucher programs enroll over 25,000 students in Milwaukee, Cleveland, and Florida. All of these programs are restricted to low-income or otherwise disadvantaged children.

The oldest program, established in Milwaukee in 1990 at the urging of local black leaders and then Gov. Tommy Thompson, was originally restricted to secular private schools and to fewer than 1,000 students. Then, in 1998, the Wisconsin Supreme Court ruled constitutional a much larger program that allowed students to attend religious schools as well. In 2002–03, over 11,000 students, more than 15 percent of the eligible population, were receiving vouchers up to $5,783, making it the country's largest and most firmly established voucher program.

The Cleveland program, enacted in 1996, was of lesser significance until the Supreme Court made it famous. Before the decision ruling it constitutional, vouchers amounted to no more than $2,250 and were limited to approximately 4,000 students. After the Supreme Court decision, the number of students increased to over 5,000, and the amount of the voucher, as of fall 2003, was as high as $2,700.

The initial Florida program, established in 1999 after Gov. Jeb Bush had campaigned on the issue, initially had less than 100 students but is poised to become somewhat larger. Here, vouchers are offered to low-income students attending failing public schools. Initially, only two schools in Pensacola were said to be failing, but in 2002, ten more joined their ranks. A second Florida program, which offers vouchers to students eligible for special education services, has received less attention but is perhaps more significant. In 2002–03, over 8,000 of Florida's special education students were enrolled in nearly 500 private schools.

In addition to these publicly funded voucher programs, there are in the United States numerous privately funded scholarship programs that operate much like school voucher programs. All these programs limit the scholarships to students from low-income families. They allow the parents to pick the private school of their choice, but they pay approximately half the tuition for more than 60,000 students. The largest program, operated by the Children's Scholarship Fund, offered 40,000 vouchers to students nationwide; over a half million students applied, and a lottery was used to select the winners. In New York City, Washington, D.C., and Dayton, Ohio, lotteries were also used to select winners from a large number of other applicants.

In other words, a variety of privately and publicly funded voucher programs are in operation. We can now look to see whether a program that works in theory also works in practice. To put the theory to a careful test, we shall report results from three randomized field trials similar to clinical (pill–placebo) trials conducted in medical research, generally regarded as the gold standard of scientific research. Both the nationwide Children's Scholarship Fund program as well as the programs in New York City, Washington, D.C. and Dayton, Ohio, were conducted in a manner that allowed for a randomized field trial because the voucher recipients were chosen by lot. This enabled the evaluation team to compare those students and families who won the lottery with a control group of students and parents who requested but did not receive a voucher and remained in public school. The two groups of students—and their families—are, on average, similar in all respects because the only difference between them is that one group won the lottery while the other did not. This apple-to-apple comparison allowed for a rigorous testing of a variety of propositions drawn from classical economic theory.

Proposition 1: Market-Based Schools Tailor Services to Consumer Demand

Markets enhance the efficiency with which goods and services are provided simply by giving consumers services they prefer. Producers have an incentive to create products for which there is a demand and to abandon those that have little appeal to consumers. If men become bored with bell-bottom trousers, retailers will not stock them. Production will instead shift to blue denim cutoffs that strike a popular chord. So it is with schools. Private schools that survive only if parents choose them are more likely to provide goods and services that are in high demand; public schools, funded by taxpayer dollars, are less likely to be so responsive.

One can test this proposition quite simply by looking at some basic characteristics of a school. Parent surveys have long shown that parents prefer small schools, K–8 (rather than middle or junior high) schools, smaller classes, and more orderly environments. If market theory is correct, then we should expect private schools to match parent preferences more closely than public schools do.

School Size. The ideal school size has never been identified. Scholars have never been able to show convincingly whether students learn more in big schools or small ones. Some studies indicate big schools are to be preferred; others report opposite conclusions. Most studies show that school size makes no significant difference at all. Nor do educational professionals agree on the optimal size. According to some, large schools permit a varied curricula, social experimentation, student diversity, and economies of scale. But others say the intimate atmosphere of a small school is crucial for effective learning.

But if scholars and educational professionals find it difficult to reach a consensus, most parents have drawn their own conclusions. They like small schools. All else equal, they will take a small school over a big one.

Well aware of parental preference, private schools, operating in a marketplace, give parents the size of school they prefer. If parents receive a voucher, they will be able to place their child in a smaller school. Parents of children participating in the evaluations in New York, Washington, D.C., and Dayton said their son or daughter, if in private school, had an average of 278 schoolmates. By contrast, students who remained in public school had, on average, 450 fellow students.

Age Structure of School. Should students attend schools that have a broad or narrow age range? Should children remain in the same school through eighth or ninth grade? Or should they change to a middle school after grade four? Or to a junior high school after grade six? Traditional educators favor schools with grades K–8, as in the days of the little red schoolhouse. But in response to studies by progressive educators, many school districts today have established middle schools and junior high schools.

Drawing on the tenets of classical economic theory, one expects private schools not to follow suit. Once they have recruited a student customer, they are likely to want to keep the child for as many years as possible. They will thus try to keep older students at their school for as many grades as feasible. And it is likely, though not certain, that most parents prefer elementary schools to middle schools and junior high schools, if simply to avoid the anxiety of changing schools but also, perhaps, to avoid schools that must deal wholesale with the problems associated with puberty and adolescence.

Our findings are consistent with these expectations. In our study, we found fewer students moving from one private school to another simply because they had "graduated" from an elementary school to enter a middle or junior high school. In New York City, for example, the percentage of young students changing schools just because they were "graduating" was 15 percent higher if the child was in a public school.[2] In short, private school students are more likely to stay in the same school for a longer sequence of grades.

Class Size. Among scholars, there is no more consensus on class size than on the optimum size of or the appropriate age structure for a school. Some econometric studies show that students perform better in smaller classes. Others show that the size of the class, within the fifteen- to thirty-student range, makes little difference. Still other studies suggest that class size makes a difference only if teachers are of low quality.

But if scholars cannot agree, parents—and students—can. The demand for smaller classes is an educational universal. Only those who have to pay for the smaller class demur, simply because class-size reduction is one of the most expensive of all educational innovations.

Despite the cost, private schools are more likely to respond to the market demand than public schools. In our study, there were, on average, twenty students per class in the private schools attended by participants in the study, as compared with twenty-three students per class in the public schools attended by those in the control group.

The fact that students attending private schools sat in smaller classes is, in fact, little short of astonishing, given the fact that expenditures per pupil are much higher in the public sector (see below). Reducing class size is an expensive proposition because smaller classes require the recruitment of more teachers, raising the personnel costs at the school. Private schools nonetheless make a special effort to keep their classes as small as fiscally feasible because market demand for this characteristic of educational services is particularly strong.

Discipline. Educational professionals disagree over the appropriate learning climate a school should seek to create. Old-fashioned educators generally imposed strict rules. But progressive theorists say a more relaxed climate that allows students to pursue their own interests in a flexible manner provides greater opportunities for self-expression. Their position has been reinforced by civil libertarians who have sought to protect student rights.

But even though the appropriate balance between school order and individual creativity and self-expression is hotly contested among educational and legal theorists, most parents expect an orderly, disciplined school, where learning can go forward unimpeded by rowdiness and conflict. It is difficult to imagine a private school surviving if its disciplinary climate is problematic. Low-income parents are unlikely to pay tuition to a school that is known to have serious problems with cheating, fighting, truancy, or racial conflict. But when students are assigned to a public school on the basis of residential location, families will have no choice but to send their child to that school, despite the severity of its discipline problems. Given this clear market demand, private schools can be expected to respond by placing a greater priority on maintaining order in school than public schools do.

That is precisely what we found. Parents were asked to rate how serious a problem at their child's school were each of the following: fighting, cheating, property destruction, truancy, tardiness, and racial conflict. In each case, the problem at the school was less for those children attending a private school. For example, only 32 percent of the private school parents in the three cities said fighting was a serious problem, while 63 percent of the public school parents said it was. Property destruction was said to be a serious problem by just 22 percent of the private school parents, but by as much as 42 percent of the public school ones. Racial conflict was a problem for 22 percent of the students at private schools, compared with 34 percent of those in public school. In interpreting these findings, it is important to keep in mind that the children and families are similar (except that one group won the voucher lottery), so the differences between the public and private schools must be attributed to the learning environment at the school, not to family characteristics.

In sum, market-based schools are more likely to give customers preferred services—smaller schools with broad age ranges, smaller classes, and more orderly educational environments.

Proposition 2: Market-Based Schools Will Communicate with Customers

Educational theorists differ as to the appropriate role that parents should play in their child's education. Although many believe that parents should be involved in their child's educational experiences, others wonder whether excessive involvement will encourage parental interference in the educational process or introduce inequities, as better educated, higher income parents seek special advantages for their child. Many school boards, for example, restrict parents from participating in fund-raisers for their child's school on the grounds that it gives the children at these schools advantages not shared by students elsewhere, where parents may be less motivated.

But if educational theory is uncertain as to the desirability of parental involvement in the work of the school, classical economic theorists expect private schools to ignore any doubts on this score. According to classical theory, private firms are expected to search for ways of better communicating with and involving their customers with their product—simply in order to maintain their consumer base. Retailers expend vast sums acquiring and maintaining information on those who have been customers in the past—on the reasonable assumption that these are precisely the individuals most likely to make similar purchases in the future. Once a family buys a telephone from Circuit City or Radio Shack, the company routinely duns them with information on their latest gadgets. High school seniors who express the slightest interest in a private college will soon discover their mailboxes full of campus photographs taken at the loveliest time of the year.

If classical theory is correct, then private schools will put aside any doubts about equity or excessive parental involvement and develop techniques for involving parents in the work of the school. For children matriculated at a school, retention will become a major priority, in part out of a concern for the well-being of the child, but, according to market theory, also because continuing revenue flows from tuition are essential to the school's survival. Schools will develop regular channels of communication with parents so as to ensure their engagement in the life of the school—in part because most educators believe parents should be involved in their child's education, but also, classical theory says, because engagement reinforces commitment and retention.

For public schools, retention of students and engagement of parents are less critical. Schooling is compulsory until the age of sixteen; funding comes from the taxpayer, not from tuition; and most school officials enjoy job protection. Public school officials will thus have fewer market incentives and will place higher priority on the need to protect the school from excessive parental involvement.

Homework. The issue arises even when it comes to assigning homework. Many educators urge teachers to exercise caution when assigning homework. If schools expect students to work on their studies at home on a daily basis, then parents are given routine opportunities to influence—even interfere with—the learning process. The better educated families can use this as a vehicle to give their child special advantages.

Private schools pay little attention to such advice, however. More regularly than public schools, they assign homework to students, and when they do, the homework is regarded by the parents as more appropriate. In the three cities, 72 percent of the private school parents reported that their child had more than one hour of homework per day, compared with 56 percent of the public school parents. Ninety percent of the private school parents said the homework was at the appropriate level of difficulty, compared with 72 percent of the public school parents.

School Communications. Private schools also communicate more frequently with parents in other ways. Private school parents were more likely than the control group of public school families to say they receive a newsletter from the school, participate in instruction, are notified of disruptive behavior the first time it happens, receive regular notes from the teacher, speak to classes about their job, are kept regularly informed about student grades, and attend open houses at the school. They are also more likely to be asked to participate in fund-raising activities.

In interpreting these findings, it is important, once again, to remember that the groups of parents whose children attend public and private schools in this study were similar because it was just random chance—a lottery—that determined whether or not they received a voucher opportunity. The enhanced parental engagement with the school was not due to special qualities of the private school parents; rather, it was due to the greater efforts by the school to involve these parents in the educational life of the child. Classical theory suggests that these schools have a strong interest in doing so.

Retention Rates. How do these efforts by private schools to maintain communications with families affect their retention rates? Classical theory expects higher turnover in the private than in the public sector simply because, in the private sector, parents are paying for the child's education. And if children are going to public school, compulsory education laws ensure that they remain in school. Furthermore, private firms wish to keep only those consumers who contribute to profits. They do not want customers who fight, steal, disrupt the business environment, and loiter for long periods of time without purchasing a product. None of these folks are good for business. Similarly, private schools can be expected to ask students to leave if they do not concentrate on their studies and comport themselves appropriately. Meanwhile, public schools are expected by law to provide for the schooling of all those living within their jurisdiction. One therefore expects higher rates of suspension, expulsion, and turnover in private than in public schools.

Surprisingly, classical theory, for once, seems to fail us. Private and public school differences are less than these considerations suggest. When parents were asked whether their child had changed schools during the school year or anticipated a change over the summer, we found no significant differences between private and public school parents. Although turnover rates for this low-income, inner-city population were high in both sectors, there was little difference between them. Apparently, a high degree of residential mobility leads to significant turnover in the public sector, one that is roughly equivalent to that in the private sector. We also did not find, in most cases, systematic differences in student suspension rates. Generally speaking, the likelihood that a child would be suspended varied between 5 and 10 percent in both sectors. However, among older students in Washington, D.C., we discerned higher suspension rates in the private sector. These students entered private schools with vouchers after having attended public schools for several years, and it was not clear that they had adjusted easily to private-sector expectations. Nonetheless, all the evidence, taken together, reveals a greater capacity to retain low-income students in private schools than classical theory might, at first glance, lead one to expect.

There are a couple of ways of explaining the anomaly. For one thing, the schools attended by these low-income voucher recipients were themselves low-tuition schools that often were in need of additional students. Efforts to maintain enrollment may have been particularly intense. Second, students may quickly adapt to the expectations of a school if it becomes clear that they will be suspended or expelled. Just as it takes but one rotten apple to spoil a barrel, so the barrel can be preserved simply by tossing out the one bad apple. Suspension, expulsion, and turnover rates may rise in the public sector simply because students realize that attendance at the school is a matter of right. Private schools tell students from the very beginning that continuation at the school depends upon conformity to school norms.

Proposition 3: Choice Breeds Happiness

Many professional educators worry about giving parents a choice of school. If parents have choice, they may select a school for what are thought to be wrong reasons—religious affiliation, racial composition, athletic facilities, convenience, or simply the school friends are attending. They also fear the degree of educational stratification that may accompany systems of educational choice.

But if educators worry about choice, classical economic theory celebrates it. For one thing, customers are expected to be happier if they have a choice. Few propositions drawn from classical economic theory are as widely accepted as this one. Tell a customer they have no choice of doctors and they will complain bitterly about the one they have. Allow them to choose freely among medical professionals and their satisfaction levels rise.

Not everyone agrees as to just why choice breeds happiness, however. Some say that satisfaction levels, as reported in surveys, are artificially inflated because consumers hate to admit a mistake. The "lemon" one purchased from the used car dealer has a marvelous tinted window, reason enough to purchase it. The sofa is the right length, even if uncomfortable. But self-delusion has its limits. The longer one has the product, the less likely one is to ignore its deficiencies. Sooner or later, the lemon will be sold and the couch replaced.

Classical theory therefore expects to find, initially, higher levels of parental satisfaction with private schools, but it also expects these satisfaction levels to attenuate with time. What may seem to be a great new world for one's child in the first instance may not prove to be as wonderful an opportunity as the years unfold. But if some decline is to be expected, the rate of decline is dependent upon product quality. If the used car proves itself, satisfaction levels could persist for years to come.

In the evaluations of the three voucher programs in New York, Washington, and Dayton, parents were asked about their satisfaction with a wide range of school characteristics, including what was taught, teacher skill, the quality of the academic program, school discipline, school safety, student respect for teachers, class size, clarity of school goals, parental involvement, and other characteristics. At the end of the first year, parents in private schools expressed much higher levels of satisfaction. For example, 54 percent of the private school parents expressed a high level of satisfaction with the quality of the academic program, compared with 15 percent of the public school parents. For school discipline, the percentages were 53 percent and 15 percent, respectively. The pattern remained much the same for many of the other characteristics.

Because the responses to many different questions fell into a common pattern, it was possible to construct an overall satisfaction scale. Differences on this scale at the end of the first year were very large, 0.92 standard deviations, for the three cities combined. Very seldom does one find differences this large between two groups participating in a randomized field trial.

But did this very large difference in satisfaction levels persist over time, or did it sharply attenuate? We were able to track this most carefully in New York City, where we obtained satisfaction reports from parents in each of three years. At the end of the first year, satisfaction levels were 1.01 standard deviations higher among the private school parents; at the end of two years, it climbed slightly to 1.05 standard deviations; by year three, it had fallen slightly to 0.94 standard deviations. In short, consumer satisfaction with vouchers was real, not ephemeral. Choice breeds satisfaction in more than just the very short run.

Proposition 4: Market-Based Schools Are More Productive

Educators worry about the educational productivity of market-based schooling. Private schools with a religious affiliation may place a higher premium on maintaining the child's religious identity than in providing them with an education. For-profit schools may skim profits by providing the most mechanical educational experience.

Classical economic theorists think otherwise. Markets stimulate productivity, says classical economic theory, not only by better matching goods and services to consumer preferences, but also by finding more efficient ways of producing these items at higher quality. So rapid is technological innovation in the computer industry that PCs today have greater computational capacities at lower costs than those available just a year or so ago.

Similar efficiency gains are unknown to modern American public education. Here the costs—in real 2002 dollars—have climbed steadily over the past half century, rising from $3,500 per pupil in 1960 to nearly $9,500 per pupil in 2000, a near threefold increase. Despite this increase in expenditures, student performances, as measured by standardized tests, have barely budged. Admittedly, test scores are not the only item to be measured in an overall assessment of school productivity, but they certainly are among the most important. If a near threefold increase in expenditure yields no gains in a key educational outcome, certainly there are severe signs of diminishing productivity. Indeed, we know of no other major sector of the American economy that has become so markedly less productive over this period of time.[3]

But does school choice increase productivity, either by raising student performance or by reducing school costs? We were able to obtain a fair comparison of educational costs in public and Catholic schools in New York City because both systems made available to us financial records that facilitated a more considered comparison than is usually possible. To make the comparison fair, we excluded from public school costs the items that were probably not being provided by Catholic schools, including monies spent on transportation, special education, school lunches, other ancillary services, and all the costs of the administrative staff at the city, borough, and district levels. All these deductions constituted 40 percent of the total cost of public schools in New York City. The remaining public school costs in 1998 were still $5,000 per pupil, more than twice the $2,400 per pupil cost of Catholic schooling in the city.

Despite this resource gap, Hispanic students attending private schools did equally well as their public school counterparts, and African American students did strikingly better. After three years, private school African American students were performing at a level that was nearly two grade levels higher than the control group remaining in public schools. In short, private schools, with half the resources, did equally well at providing educational opportunity for Hispanic students and considerably better for African American ones. Once again, these differences cannot be attributed to higher initial capacity or commitment on the part of student or family because the two groups of students were originally similar, save for the fact that the one group had won the lottery.

Proposition 5: The Characteristics of Both Public and Private Schools Affect Voucher Usage

According to classical economic theory, both push and pull factors are likely to affect voucher usage. Families will be pushed away from public schools, if they find them unsatisfactory. And they will be drawn toward private schools, if they have qualities families find especially appealing. However, they will remain in private schools only if they remain satisfied with the new educational opportunity.

The decision to use a voucher can be broken into the following three steps: (1) applying for a voucher; (2) using a voucher, if offered one; (3) remaining in a private school over time. Each step requires a greater commitment than the previous one, especially when vouchers pay only about half the cost of attending a private school (as was the case in the situations examined here).

The process of obtaining and using a voucher can be usefully compared to the processes of courtship and marriage with which most are familiar. The initial decision to date requires little commitment. If sufficiently unhappy, the love-starved may agree even to a blind or computer-generated date. Factors explaining decisions at this point are more likely to be "push" considerations, such as prolonged loneliness, the collapse of a previous love affair, or a divorce. Agreeing to marriage is another matter, one that must be taken seriously by both parties. Here the pull factors of the potential mate are more likely to be critical. And, of course, the marriage persists only if the relationship is successful.

So it is with vouchers. Each step—from initial expression of interest to the decision to matriculate at a specific school to retention at that school—requires a greater commitment on the part of both the parent and the private school the child is attending. Considerations that induce voucher applications are not always the same as those that lead families to use them, when offered the opportunity, or to keep families within voucher programs over time.

The best information on the first stage of the process, the application for a voucher, comes from an evaluation of the nationwide scholarship program administered by the Children's Scholarship Fund. In this case, my colleagues and I were able to compare low-income applicants with all low-income families eligible for participation in the voucher program.[4]

Push factors were important at the applicant stage. Those who applied were less likely to be satisfied with the public school their child was currently attending. Only 24 percent said they were "very satisfied" with the academic quality of the school, compared with 38 percent of all eligible parents. Satisfaction with public schools may also help explain why vouchers were also much more attractive to African American families than to either white or Hispanic families. Forty-nine percent of the applicants were African American, compared with just 26 percent of the eligible population. Presumably, public schools attended by African American students are particularly problematic.

But pull factors were also important. Families were more likely to apply for a voucher when they were actively engaged in religious life. Since most private schools have a religious affiliation, it is very likely that this religious dimension was something these families were seeking. Otherwise, differences between applicants and eligibles were modest. Applicants were only slightly more likely to live in two-parent households or to have mothers who were college educated. Even those with disabilities were as likely to apply as those who were not. However, applicant families were more likely to have lived in their current residence for two or more years, a sign that voucher applicants were better embedded in community networks than eligible families more generally.

Pull factors become especially important at the second stage of the voucher utilization process, the point at which lottery winners must decide whether to use the voucher offered them. At this point, a critical pull factor is the sheer availability of a private school. Thus, usage rates were higher in those metropolitan areas where the private school share of the market was the greater. Another indication that families were being drawn to the private sector is the fact that those regularly engaged with religious institutions, especially if Catholic, were more likely to use the voucher. Since over two-thirds of private schools are Catholic, the availability of private schools to active members of this religious faith gave these families a special opportunity. Financial issues also seem important, inasmuch as family members with more children were less likely to take up the opportunity. For low-income families, placing several children in a private school may have been too taxing, especially since in this program the voucher usually covered only about half the cost.

Finally, evidence with respect to differences in ethnic response at this stage of the process is mixed. In the national Children's Scholarship Fund evaluation, African American families were much *less* likely to use a voucher when offered the opportunity. But in New York City, they were much *more* likely to use vouchers, if offered.[5] These quite opposite effects remain large even after many other factors are taken into account in the analysis. Nor is there reason to question the quality of the data in either case. The inconsistency of the findings from the two evaluations may be reconciled by considering a key pull factor—the availability of private schools in African American neighborhoods. In New York, private schools may have been readily available to African American students, in part because many Catholic schools remain in the New York neighborhoods where African Americans live. The Catholic immigrant groups that built the schools have left these neighborhoods, but the well-established Catholic archdiocese in the city has made strenuous efforts to keep the schools intact. This is probably less true nationwide. Private schools, Catholic or not, may be scarce in neighborhoods with a high concentration of African American families.

The New York evaluation also provides information concerning those who are willing to remain in private school over a three-year period. As might be expected, satisfaction with the private school critically affects the likelihood of leaving the program. Also significant is the match between the religious affiliation of the family and the school, a sign, once again, that preference for a particular kind of educational experience is important to voucher users. Finally, African American students are less likely to remain in private school than are students from other ethnic groups.

In short, both push and pull factors affect voucher usage. Families are attracted to a voucher program if they are dissatisfied with public schools and/or they seek special qualities (such as religious engagement) from a private school. But they are unlikely to continue to use a private school if they become dissatisfied with its quality.

Proposition 6: Public Schools Will Respond to Competition, Perhaps

If classical economic theory is correct, then public schools, confronted by the possibility that they could lose substantial numbers of students to competing schools within the community, may be expected to respond by reaching out more effectively to those they are serving.

In the randomized field trials we conducted, the number of voucher students was too small for their presence to have any discernable impact on the public schools in these cities. But in Milwaukee, voucher students constitute over 10 percent of the student population whose education is publicly financed. Another 10 percent of the students attend charter schools, which also provide families with a choice of school. Substantial school choice has been available to families since 1998, providing the best setting in which to identify how vouchers impact public schools in the vicinity.

Early research on Milwaukee suggests that vouchers are having an impact on the public schools, albeit slowly. Relying on evidence collected in 1999, only one year after the expanded program had begun, American Enterprise Institute scholar Frederick Hess concluded that public schools had few incentives to respond to the competition—in part because their revenues and the job opportunities of school employees were protected from the competition. At least in the first few years, the schools seemed to be making little more than symbolic responses to the competition.[6] But other evidence is more encouraging. Harvard economist Caroline Minter Hoxby found signs that public school test scores rose more rapidly in those Wisconsin public schools that were impacted by vouchers.[7] Even the threat of a voucher can have a positive effect on test scores. Research by Manhattan Institute scholar Jay Greene shows that when public schools were in danger of failing twice on the statewide Florida exam, making their students eligible for vouchers, these public schools made special efforts to avoid failure.[8]

Despite these positive early signs, one cannot expect rapid transformation of public schools, even if voucher programs should expand, simply because public financing arrangements are often designed to protect public schools against competition. Although financial arrangements vary from one state to the next, on average, nationwide, 49 percent of the revenue for public elementary and secondary schools comes from state governments, while 44 percent is collected from local sources and the balance received in grants from the federal government. Most of the revenue school districts get from state governments is distributed on a "follow the child" principle. The more students in a district, the more money it receives from the state. If a child moves to another district, the state money follows the child. Local revenue, most of which comes from the local property tax, stays at home, no matter where the child goes. As a result, the amount of money the district has per pupil actually increases if a district suffers a net loss of students, simply because local revenues can now be spread over fewer pupils.

The voucher programs in Milwaukee, Cleveland, Florida, and Colorado have all been designed to protect public schools from serious financial problems when students accept vouchers. The state money follows the child, but the local revenue stays behind in local public schools, which means that more money is available per pupil. In Milwaukee, per-pupil expenditures for public school children increased (in real dollars) by 22 percent between 1990, when the first small voucher program began, and 1999, when vouchers were prevalent. The rise in expenditures was from $7,559 to $9,036. Not all of the increase was a direct result of the voucher program, but the example shows that public schools do not necessarily suffer financially when voucher programs are put into effect.

In short, public schools thus far have few financial incentives to respond to voucher competition.

Proposition 7: Economic Logic Does Not Necessarily Translate into Political Logic

School vouchers in practice seem to operate much as the Friedmans have long suggested they would work in theory. When theoretically well-grounded innovations prove successful in practice, one ordinarily expects a fairly rapid diffusion of the innovation. According to classical economic theory, followers will adopt the innovations of industry leaders simply in order to survive the competitive threat.

Such a response is less likely, however, when vested interests adversely affected by the innovation can use government authority to keep the innovation from spreading. In the early seventeenth century, the watermen of London sought to keep wagons and coaches from appearing on the city streets. A perceptive architectural historian tells the story in this way:

> One gets an impression of the importance of the [Thames] river traffic on hearing that in 1613 the number of the watermen and their families

amounted to 40,000 in a city whose entire population hardly exceeds 200,000. By means of propaganda, they made war on all other methods of transport, by wagon or by coach, but it was of no use. In 1601 they succeeded in getting a Bill passed in the House of Commons "to restrain the excessive and superfluous use of coaches." This was, however, stopped by the House of Lords.[9]

While the watermen failed in this instance, they regularly impeded the advancement of land transport in the decades to follow. Similarly, throughout much of the twentieth century, American railroads used their access to the Interstate Commerce Commission to protect themselves from the trucking industry. Today, pharmaceutical companies routinely fight the deployment of generics as an infringement on their patents. In short, government authority to regulate is often used to protect producers from competition.

Public schools, as traditionally organized, are no less well positioned to protect their interests than were London's seventeenth century watermen. Much like London's river traffic industry, the educational industry is today very large, constituting no less than 5 percent of the American economy. Most Americans once attended public schools themselves, and, as a result, their affection for this institution, no matter how aging and sluggish, is deep and abiding. The industry's political flank is well protected by two major unions, the National Education Association and the American Federation of Teachers, which are among the most active organizations in national, state, and local politics. In local school board elections, teachers vote with a frequency unrivaled by ordinary citizens — especially if they live and work in the same district.[10] Fighting the spread of school vouchers is a top union priority. When doing so, unions can invoke the public school as the symbol of democracy and vouchers as an unconstitutional threat to the unity of the American people.

THE CONSTITUTION AND BALKANIZATION

Ever since the voucher concept was first enunciated by the Friedmans, its constitutionality has been questioned by those who said it violated the establishment of religion clause of the U.S. Constitution's First Amendment. But in 2002, a five-member majority of the Supreme Court found, in the case of *Zelman v. Simmons-Harris*, that the Cleveland school voucher program was constitutional. The court declared that the program did not violate the establishment clause, as plaintiffs had argued, because it allowed parents a choice among both religious and secular schools. There was no discrimination either in favor of or against religion.

But even though school vouchers have passed this crucial constitutional test, many have argued that they would prove divisive in a pluralist society with multiple religious traditions. In his dissent from the majority opinion in *Zelman*,

Justice Stephen Breyer saw the decision as risking a "struggle of sect against sect." And Justice John Stevens said he had reached his decision by reflecting on the "decisions of neighbors in the Balkans, Northern Ireland, and the Middle East to mistrust one another. . . . [With this decision] we increase the risk of religious strife and weaken the foundation of our democracy."

These dissents echo the concerns of many distressed by the worldwide rise in fundamentalist religious conviction, worries that have intensified since 9/11. But though the concerns are genuine enough, it's hardly clear that government-controlled indoctrination of young people is the best tool for conquering intolerance. On the contrary, this strategy proved counterproductive in many parts of the former Soviet Union. Historically, the United States has achieved religious peace not by imposing a common culture but by ensuring that all creeds, even those judged as dangerous by the enlightened, have equal access to democratic processes.

Of course, religious conflict is part and parcel of American political history. In the late nineteenth century, many objected to the establishment of Catholic schools. Indeed, anti-immigrant sentiment was so strong that amendments to state constitutions were enacted that seemed to forbid aid to religious schools. Many of these provisions are the so-called Blaine amendments, dating to the nineteenth century, when James Blaine, a senator from Maine and a Republican presidential candidate, sought to win the anti-immigrant vote by campaigning to deny public funds to Catholic schools. (Blaine is perhaps most famous for tolerating a description of Democrats as the party of "Rum, Romanism, and Rebellion.") In its classic version, the Blaine amendment read as follows:

> No money raised by taxation for the support of public schools, or derived from any public fund therefore, nor any public lands devoted thereto, shall ever be under the control of any religious sect; nor shall any money so raised or lands so devoted be divided between religious sects or denominations.

Blaine-like clauses in state constitutions are being invoked by those seeking to forestall voucher initiatives. In a number of cases, state courts have interpreted these clauses to mean nothing more than what the Supreme Court defines as the meaning of the establishment clause of the First Amendment. If this view prevails in state courts, then vouchers do not violate these state constitutional clauses now that they have been found constitutional by the U.S. Supreme Court. And if the Blaine amendments are invoked as a basis for finding vouchers in violation of state constitutions, the Supreme Court may eventually be asked to decide whether, on account of their nativist and anti-Catholic origins, these Blaine amendments—and their derivatives—are themselves unconstitutional.[11]

The controversies over religion seem more heated in the political and legal world than in the classroom, however. While exceptional cases can always be identified, there is little evidence that religious schools typically teach intolerance. Indeed, careful studies have shown that students educated in Catholic schools are both more engaged in political and community life and more tolerant of others than public school students. After enduring harsh criticism from critics in a Protestant-dominated America, Catholic schools took special pains to teach democratic values.[12] The more recently established Christian, Orthodox Jewish, and Muslim schools can be expected to make similar attempts to prove they, too, can create good citizens.

As Justice Sandra Day O'Connor pointed out in her concurring opinion, if Justices Breyer's and Stevens' fears were real, we'd know it already. She showed that taxpayer dollars flow to religious institutions in multiple ways—through Pell Grants to sectarian colleges and universities; via child care programs, in which churches, synagogues, and other religious institutions may participate; and through direct aid to parochial schools of computers and other instructional materials. If thriving religious institutions create a Balkanized country, she seems to say, this would already have happened.

Nor, say voucher proponents, have public schools eliminated social divisions. As Justice Clarence Thomas argued in his concurring opinion, "The failure to provide education to poor urban children perpetuates a vicious cycle of poverty, dependence, criminality and alienation that continues for the remainder of their lives. If society cannot end racial discrimination, at least it can arm minorities with the education to defend themselves from some of discrimination's effects." In other words, vouchers may help heal, not intensify, the country's most serious social division.

NOTES

[1] Unless otherwise indicated, the information in this paper is taken from William Howell and Paul E. Peterson, with Patrick Wolf and David Campbell, *The Education Gap* (Brookings, 2002), 92. The research from randomized field trials summarized in this paper is reported in full in this monograph.

[2] David Myers, Paul E. Peterson, Daniel Mayer, Julia Chou, and William Howell, "School Choice in New York City After Two Years," Program on Education Policy and Governance, Working Paper No. 00-17 (August 2000), Table 17.

[3] For these facts and other related information on the productivity of the American school system, see Paul E. Peterson, ed., *Our Schools and Our Future: Are We Still at Risk?* (Hoover, 2003), Ch. 2–3.

[4] Paul E. Peterson, David E. Campbell, and Martin R. West, "Who Chooses? Who Loses? Participation in a National School Voucher Program," in Paul T. Hill, ed., *Choice with Equity* (Hoover, 2002), 51–85.

[5] William G. Howell, "Dynamic Selection Effects in School Voucher Programs," *Journal of the American Association of Public Policy and Management*, forthcoming.

[6] Frederick M. Hess, "The Work Ahead," *Education Next* I (Winter 2001), 8–13; Frederick M. Hess, *Revolution at the Margins* (Brookings, 2002).

[7] Caroline Minter Hoxby, "Rising Tide," *Education Next* I (Winter 2001), 69–74.

[8] Jay P. Greene and Marcus Winters, "When Schools Compete: The Effects of Vouchers on Public School Achievement," Education Working Paper No. 2, Center for Civic Innovation at the Manhattan Institute (2004).

[9] Steen Eiler Rasmussen, *London: The Unique City* (Penguin, 1961), 116–17.

[10] Terry Moe, "Political Control and the Power of the Agent," Hoover Institution, Stanford University, 2003 (Unpublished paper).

[11] A related issue is being considered by the Supreme Court during its 2003–04 term. See James E. Ryan, "The Neutrality Principle," *Education Next* III (Fall 2003), 28–35.

[12] For citations, see Howell and Peterson, with Wolf and Campbell (2002), 130–32.

Session 2

The Property Rights Path to Sustainable Development
Terry L. Anderson and Laura E. Huggins

Economic Freedom and Environmental Quality
Richard L. Stroup

The Property Rights Path to Sustainable Development

Terry L. Anderson and Laura E. Huggins

> *You can't have a free society without private property.*
> —Milton Friedman

Sustainable development has become the byword of environmental policy. The term has been around for about thirty years but has only recently become popular (see International Institute for Sustainable Development 2002). The basic notion is that current consumption of natural resources, including air and water for waste disposal, is reducing the stock or quality of those resources so that future generations will have less. If it is true that there are finite resource stocks, consumption today will preclude sustained consumption in the future.

In this sense, sustainable development dates back to the eighteenth-century writings of Reverend Thomas Malthus, who believed that the decline of living conditions in nineteenth-century England was due to the inability of resources to keep up with the rising human population, and more recently to the "limits to growth" theory promoted by the Club of Rome in the 1970s. Armed with giant computers, this group predicted precise years when we would reach our limits. Their predictions of disaster for humankind called for regulations restricting natural-resource-depleting economic and technological progress.

This gloom and doom theory has been resurrected under the guise of sustainable development, calling for changes in virtually every aspect of our consumption and production. Starting in the late 1980s, environmentalists and government officials began using the term "sustainable development" when discussing environmental policy. For example, the seminal United Nations Bruntland Report (Bruntland 1987, 9) claimed that "sustainable development can only be pursued if population size and growth are in harmony with the changing productive potential of the ecosystem." A paper by the U.S. Department of Housing and Urban Development (1995, 2) declared that "humanity's collective imperative now is to shift modern society rapidly onto a sustainable path or have it dissolve of its own ecologically unsustainable doings."

More recently the interpretations of the term have been broadened to include issues such as poverty, health care, and education. The Johannesburg Declaration on Sustainable Development (2002, 2) stated that "poverty eradication, changing consumption and production patterns, and protecting and managing the natural resource base for economic and social development are overarching objectives of, and essential requirements for, sustainable development."

The term sounds beguilingly simple, but it is vague and operationally vacuous. Sustainable development is most often defined as resource use that meets "the needs of the present without compromising the ability of future generations to meet their own needs"—a definition first offered by the United Nations' Bruntland Commission. Of course, nobody wants to make future generations poorer and less healthy, but this definition provides no guidance for how this result can be avoided. There is no way to know what resource use is acceptable today and no way to know what future generations may desire (see Hayward 2002). Yet because of its deceptive simplicity, sustainability is applied to anything from agricultural practices to energy use to mining. As environmental scientist Timothy O'Riordan stated, "It may only be a matter of time before the metaphor of sustainability becomes so confused as to be meaningless, certainly as a device to straddle the ideological conflicts that pervade contemporary environmentalism" (O'Riordan 1988, 29).

Implicit in the calls for sustainable development are two fundamental assumptions. The first is that we are running out of resources, thus leaving future generations with less; the second is that market processes are the cause of these depletions. We challenge both of these assumptions and argue that economic systems based on property rights and the rule of law are the best hope for humanity today to leave an endowment for humanity tomorrow.

For those familiar with the writings of Milton and Rose Friedman, there is nothing new in this conclusion. In *Free to Choose*, the Friedmans forcefully argued that political and economic freedom are inseparable and that free-market forces work better than government controls for achieving real equality, security, and prosperity.

In this paper, we build on the Friedmans' case to argue that sustainable development, if it can be defined, is only possible in a legal system where property rights are well defined, enforced, and transferable. Property rights provide the structure that encourages development, innovation, conservation, and discovery of new resources. Growth and increasing wealth through these mechanisms lead to environmental quality by raising the demand for it and by providing the wherewithal to meet these demands. In this context, economic growth is not the antithesis of sustainable development; it is the essence of it.

Therefore, how we deal with the evolution and protection of property rights and the rule of law in the future will not only determine how free and prosperous we are, but also how much environmental quality we enjoy.

THE PROVEN PATH TO SUSTAINABILITY

If sustainable development can be defined as a call to maximize human welfare over time, then Adam Smith's *Wealth of Nations* could be called "the world's first blueprint for sustainable development" (Taylor 2002, 29). According to Smith's blueprint for sustainable development, the wealth of nations depended on market processes guided by the invisible hand, which we now understand was not so invisible. Perhaps no one since Adam Smith has so eloquently made it clear that the invisible hand is really property rights and the rule of law as have Milton and Rose Friedman. Institutions such as property rights and the rule of law provide the framework within which people act and interact. They are the rules, customs, norms, and laws that remove the responsibility to calculate the effect of our actions on the rest of humanity and replace it with a responsibility to abide by simple rules that benefit society as a whole. In the words of Richard Epstein, "The government works best when it establishes the rules of the road, not when it seeks to determine the composition of the traffic" (Epstein 1995, xiii).

Sustainable institutions are those that do not prescribe an outcome for society, but allow individuals to improve their own well-being. Truly sustainable institutions provide the freedom for people to improve their world by innovating and developing. The best way to ensure that resources remain for future generations is to directly tie the well-being of people today to those resources—via decentralization and property rights. If individuals have the responsibility of caring for their welfare today, they are more likely to make decisions that will benefit their children, and their children's children (see Taylor 2003).

Modern data support the conclusion that Adam Smith's blueprint works. The Fraser Institute's *Economic Freedom of the World: 2003 Annual Report* rates and ranks 123 nations based on thirty-eight variables to conclude that the more economically free a country, the greater the level of human development enjoyed by its citizens (Gwartney, Lawson, and Emerick 2003). Figure 1 helps illustrate the basic notion that economic freedom contributes to a faster growing and more efficient economy, which translates into better and longer lives. "Freeing people economically unleashes individual drive and initiative and puts a nation on the road to economic growth," says Milton Friedman, one of the original creators of the economic freedom index. "In turn, economic growth and independence from government restrictions promote civil and political liberty" (quoted in Gwartney, Lawson, and Emerick 2003).

These findings are supported by other scholars. Seth Norton, for example, has calculated the statistical relationship between various freedom indexes and environmental improvements. His results show that institutions—especially property rights and the rule of law—are key to human well-being and environmental quality. Norton examined the role of economic institutions on human well-being by dividing a sampling of countries into groups with low, medium, and high economic freedom and the same categories for the rule of law. Table 1 contains the measures of human well-being for the various groups of countries. In all cases except water pollution, those in countries with low economic freedom are worse off than those in countries with moderate economic freedom, while in all cases those in countries with high economic freedom are better off than those in countries with medium economic freedom. A similar pattern is evident for the rule of law measures (see Norton 2003).

Theodore Panayotou (1997, 465–84) tested five indicators of general institutional quality: "respect/enforcement of contracts, efficiency of the bureaucracy, efficacy of the rule of law, extent of government corruption, and the risk of appropriation." He found that higher indexes for the institutional variable led to significant environmental quality improvements. In another study, Madhusudan Bhattarai (2000) found that civil and political liberties, the rule of law, the quality and corruption levels of government, and the security of property rights were important in explaining deforestation rates in sixty-six countries across Latin America, Asia, and Africa. Without question, institutions—especially prop-

Figure 1
Economic Freedom and Economic Growth
GDP per capita, percent growth (1992–2001)

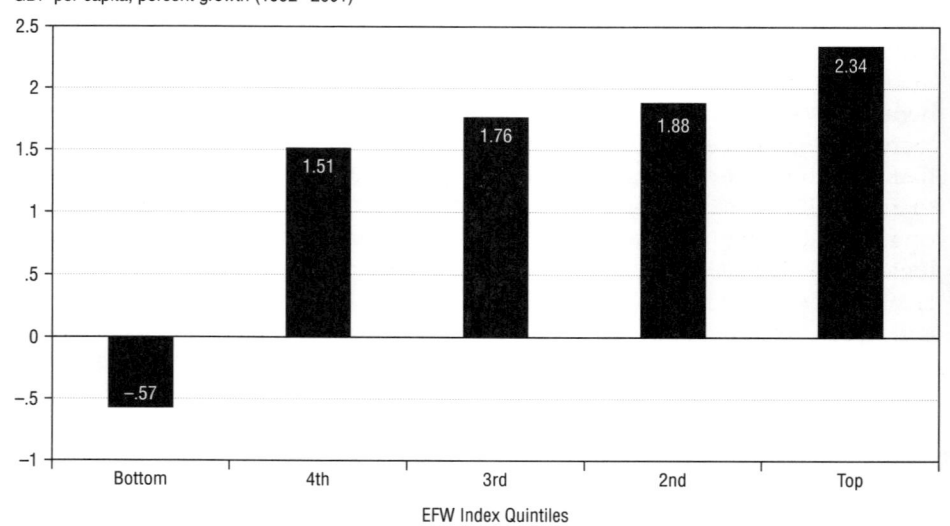

EFW Index Quintile	Value
Bottom	−.57
4th	1.51
3rd	1.76
2nd	1.88
Top	2.34

SOURCE: *Economic Freedom of the World: 2003 Annual Report.*

Table 1
Economic Institutions and Human Well-Being

	Economic Freedom			Rule of Law		
Measure of Well-Being	Low	Medium	High	Weak	Medium	Strong
Poverty index	38.1	30.5	14.5	31.8	33.0	16.4
Death by 40	29.1	19.4	7.7	19.6	21.7	10.8
Adult illiteracy	39.2	34.7	12.5	32.1	37.8	17.0
Safe water	43.3	34.7	19.5	34.8	36.2	20.1
Health service	40.5	28.5	16.8	41.3	28.0	15.2
Undernourished children	29.1	21.7	13.9	25.0	23.1	14.0
Deforestation rates	.4	1.4	−.2	1.3	.7	.3
Water pollution	.2	.2	.2	.2	.2	.2
Net savings rates	4.0	7.1	14.8	2.6	6.3	16.0
Agricultural productivity	620.3	1,011.2	6,001.6	1,178.2	1,083.6	4,552.7

SOURCES: Gwartney and Lawson (2001); Political Risk Services (1997); United Nations Development Program (1997); World Bank (2001).

erty rights and the rule of law—are key to environmental improvements (see Anderson 2003).

Policies for sustainable development that prescribe forgoing economic growth in the name of preserving future resources could stall or reverse a proven path of progress. The temptation to impose new layers of government regulation in order to prevent humans from depleting resources for future generations must be pushed aside. Consider the use of whale oil for energy in the nineteenth century. The whale population was unsustainable due to heavy hunting pressures. The near depletion of whales may have threatened the biological diversity of the planet, but the loss of whale oil as a resource did not hamper future generations from meeting their needs. More plausibly, according to Steven Hayward (2002, 4), "the use of whale oil facilitated economic development—growing wealth, incomes, occupational specialization, and technological prowess—that put humankind in a position to adopt better, more efficient, more sustainable methods of production." The demand for whale oil contributed to the development of petroleum and electricity, which were more efficient than whale oil and hence helped restore the whale population.

GROWTH UNBOUND

A popular interpretation of sustainable development presumes that environmental degradation is predominantly caused by, and therefore is the responsibility of, rich countries. People in the wealthy world consume a large proportion of the world's resources and emit too great a proportion of the world's pollution. Proponents often cling to the beliefs of Dr. Charles Birch (1976, 66), a member of the Club of Rome, who claimed that "the rich must live more simply that the poor may simply live."

Bjørn Lomborg, author of *The Skeptical Environmentalist*, set out to prove that resources are becoming more scarce and the environment is getting worse, but instead concluded that almost all environmental indicators are improving, primarily because we are wealthier, can afford cleaner technologies, and have more time and money to devote to environmental luxuries. According to Lomborg (2001, 351), "children born today—in both the industrialized world and developing countries—will live longer and be healthier, they will get more food, a better education, a higher standard of living, more leisure time and far more possibilities—without the global environment being destroyed." Furthermore, Lomborg finds positive correlations between economic growth and environmental quality. He correlates the World Bank's environmental sustainability index with per capita gross domestic product across 117 nations, concluding that "higher income in general is correlated with higher environmental sustainability" (Lomborg 2001, 32).

Similar results came from a recent Environmental Sustainability Index (ESI) developed by the joint effort of the World Economic Forum, the Yale University Center for Environmental Law and Policy, and the Columbia University Center for International Earth Science Information Network. The group measured 142 nations based on twenty indicators and sixty-eight related variables in order to place a sustainability score on each nation. On the ESI scale for 2002, Finland came in first, with a score of 73.9, and Kuwait came in last, with a score of 23.9.

The most significant finding derived from the ESI study compares each nation's ESI score with its gross domestic product (GDP) per capita and shows that a strong relationship exists between wealth and environmental quality, as seen in Figure 2. Careful analysis of the figure reveals a depiction of what economists call the environmental Kuznets curve, based after Nobel laureate Simon Kuznets. At lower levels of income, environmental quality can deteriorate as people trade off environmental quality for economic growth, but as income levels rise, the demand for environmental quality increases at a higher rate (see Yandle, Vijayaraghavan, and Bhattarai 2002).

The work of Indur Goklany adds further optimism to the potential for economic growth to be a major factor in improving environmental quality. In case after case, Goklany demonstrates that economic growth allows the developing world to enjoy higher living standards sooner than the developing world did in the past. For example, once a country such as the United States creates filters for water purification, developing countries do not have to "reinvent the wheel." They can simply acquire the new technology and improve water quality at lower levels of income (Goklany 2003).

Improvement of the environment with income growth is not automatic but depends on policies and institutions. Economic growth creates the conditions for environmental improvement by raising the demand for improved environmental quality and makes the resources available for supplying it. Whether environmental quality improvements materialize or not, when and how depend critically on government policies, social institutions, and the completeness and functioning of markets.

Institutions that promote democratic governments are a prerequisite for sustainable development and enhanced environmental quality. Where democracy dwells, constituencies for environmental protection can afford to exist— without people fearing arrest or prosecution. The democratization of thirty-plus countries in the last twenty-five years has dramatically improved the prospects for environmental protection (Desta 1999).

In the other direction, dictatorships and warlords burden people and environments in many regions of the world such as China and much of Africa. Zimbabwean president Robert Mugabe, for example, has clearly indicated that he has no intention of respecting property rights or the rule of law. His "terror teens" have brutally killed innocent people, and his "land reform" plan demands that more than 20 million of the 23.5 million acres under private ownership be sur-

Figure 2
Relationship Between Wealth and Environmental Quality

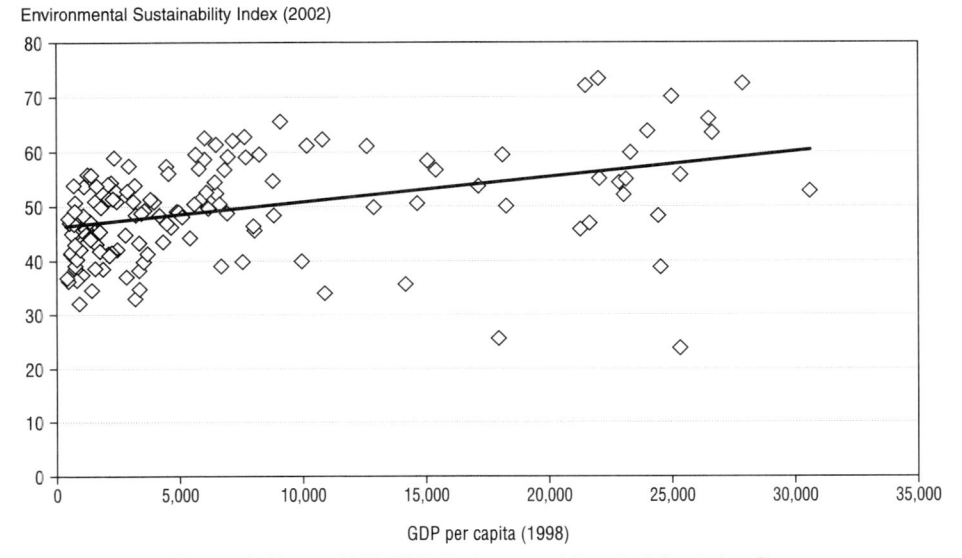

SOURCE: World Economic Forum. 2002. *2002 Environmental Sustainability Index*. Geneva, Switzerland: World Economic Forum.

rendered without compensation. Mugabe's assault on private property has also taken a toll on wildlife, for without landowners, there is no one to protect them from poachers. Before Mugabe's attack on private property, Zimbabwe had previously shown the world how to balance economic development with conservation through private and communal ownership. The CAMPFIRE program, for example, championed by the World Wide Fund for Nature, allowed local communities to manage wildlife. Hence, wildlife became an asset as villagers in communal areas profited from hunting and photo safaris. Elephant populations mushroomed and poaching plummeted. But Mugabe has duped the people of Zimbabwe into thinking that land redistribution without compensation or due process is the key to economic prosperity. In fact, sustainable development will come only from stable property rights. Unless the sanctity of private property can be reestablished in Zimbabwe, its people and its wildlife will continue to suffer.

Environmental degradation does not stem from the actions of the first world but rather from jumbled bureaucratic systems—often the result of well-meaning but misguided intervention. In particular, lack of well-defined and adequately enforceable property rights restricts economic development and stifles entrepreneurial activity in many countries. The Peruvian economist Hernando de Soto "estimates that people in the third world and in ex-communist countries hold more than $9 trillion in what he calls 'dead capital'—property that is owned informally, but not legally, and is thus incapable of forming the basis of robust economic development. He advocates the formal recognition of property rights in these countries as an indispensable prerequisite for liberal democracy" (Ponnuru 2003).

UNSUSTAINABLE REGULATIONS

The focus on center stage should be on promoting institutions that empower people both politically and economically. These institutions allow people to improve environmental quality indefinitely into the future. This stands in sharp contrast to the undying conclusion of the doomsayers for whom the environment and the plight of human beings will always be worse. Doomsayers continue to profess, as they have since Thomas Malthus, that exponential economic growth and consumption will ultimately run up against resource limits. Paul and Ann Ehrlich (1996, 11) are perhaps the gloomiest:

> Humanity is now facing a sort of slow-motion environmental Dunkirk. It remains to be seen whether civilization can avoid the perilous trap it has set for itself. Unlike the troops crowding the beach at Dunkirk, civilization's fate is in its own hands; no miraculous last-minute rescue is in the cards.... Even if humanity manages to extricate itself, it is likely that environmental events will be defining ones for our grandchildren's generation—and those events could dwarf World War II in magnitude.

Those with this mind-set often call for more government regulation to stop growth and curb consumption. For example, Klaus Töpfer (2002, 1), executive director of the United Nations Environment Program (UNEP), hopes to create a "model for international environmental governance."

Implicit in the definition and use of the term sustainable development is the acceptance that market systems fail to promote sustainability and therefore that command-and-control regulations are necessary to achieve the goal of sustainable development. Agenda 21, for instance, adopted at the 1992 Earth Summit in Rio de Janeiro, calls on governments to "create national strategies, plans, and policies" for sustainable development.

The solution offered by those who follow this interpretation is to impose top-down measures such as restrictions on the use of resources, interventions in the behavior of multinational companies, and restrictions on international trade. Yet evidence suggests the contrary.

In 1995, UNEP proposed to restrict and possibly ban twelve chemicals (persistent organic pollutants, or POPs), including DDT, considered to be damaging to human health and the environment. More recently, the Stockholm Convention (also known as the POPs Treaty) was driven by a network of NGOs and governments that called these chemicals "the dirty dozen." Because rich countries neither produce nor use any of the twelve, they would not feel the effects of such a ban. The problem is that banning chemicals or technologies, regardless of the risks they impose, does not take into account the risk of not having the technology. People in developing countries are subjected to dirty drinking water and poor sanitation, and they must cope with farming techniques that have not advanced since medieval times.

Man-made chemicals and new technologies can mean the difference between life and death for many people. Some countries such as Belize, Mozambique, and Bolivia have stopped using DDT in their malaria control programs because they fear losing the support of wealthy nations through aid. As a consequence, they have suffered a loss of human life. Several scientists have suggested that the insistence by wealthy nations to ban DDT in malaria-infected countries is a form of eco-colonialism, which may impoverish nations in the same way that imperial colonialism did in the past (Meiners and Morriss 2001). "If we are really concerned with ensuring a cleaner environment and with healthier populations," writes Richard Tren (2002), "we should concentrate on ensuring that the developing world can become wealthy. This can be achieved with open markets and liberalized trade, protection of private property and the rule of law."

If alarmists' calls are successful, we will have neither sustainable growth nor sustainable improvements in environmental quality. When given responsibility for their lives and property, individuals have generally tended to improve themselves and thus the state of the planet. As Milton and Rose Friedman (1980, 218) explain in *Free to Choose*, "If we look not at rhetoric but at reality, the air is in general far cleaner and the water safer today than one hundred years ago. The air is much cleaner and the water safer in the advanced countries of the world today than in the backward countries."

SUSTAINING PROPERTY RIGHTS

It is not resources that are too scarce, but rather the institutions that ensure freedom—political and economic systems based on secure property rights and the rule of law. Jerry Taylor (1993, 10) writes:

> The size of our resource pie is determined not by nature but by the social and economic institutions that set the boundaries of technological advance. Closed societies and economies under the heavy hand of central economic planners are doomed to live within the confines of dwindling resource bases and eventually experience the very collapse feared by the conservationists. Liberal societies, built on free markets and open inquiry, create resources and expand the possibilities of mankind.

When the Eastern Bloc countries were freed from communism, Milton Friedman called for free markets, saying, "Privatize, privatize, privatize." After more than a decade of experiments trying to create markets, however, he has modified his position, asking: "What does it mean to privatize if you do not have security of property, if you can't use property as you want to?" (Friedman 2002, xvii). Russia, for example, was able to create a democracy but no rule of law to protect private property. Corruption is prevalent, and Russia's economy has imploded. This does not trivialize its democratization efforts, but rather emphasizes that without the rule of law and protection of property, democracy by itself cannot bring automatic prosperity.

The institution of private property has had more influence than any other policy in the history of the world when it comes to enabling people to escape from poverty. As Tom Bethell (1998, 11) puts it in his book *The Noblest Triumph: Property and Prosperity Through the Ages*, "Prosperity and property are intimately connected. Exchange is the basic market activity, and when goods are not individually owned, they cannot easily be exchanged." Because of poorly defined institutions and often corrupt bureaucratic systems, a large proportion of the world's population is prevented from fully realizing the value of existing property or being able to acquire secure property rights. Hernando de Soto (2000) explained in his book, *The Mystery of Capital*, that the primary problem is that property claims in developing countries, while acknowledged within their communities, often go unrecognized by the government. As a result, these informal owners lack access to the social and economic benefits provided by secure property rights.

When well-defined and enforced property rights are absent, a "tragedy of the commons" situation often prevails. The phrase derives from the incentive to overgraze pastures that are open to all grazers (Hardin 1968). Each potential grazer has an incentive to fatten his livestock on the grass before someone else does. Open access to resources lacks two critical components that property rights systems share—exclusion and governance. Without these two components, people have little incentive to economize on the use of resources. The solution to that problem is to devolve control of resource management to individuals and local bodies, and to ensure that legal institutions support this (see Anderson and Huggins 2003).

A study in Kenya compared the rights and incentives of user groups for forest resource management in the Mt. Elgon National Park with those of users in the Mt. Elgon Forest Reserve. The authors find that in the forest reserve, which represents decentralized management, local community involvement in decisionmaking and in rule crafting and enforcement resulted in positive incentives for forest conservation. Forest conditions in the forest reserve were found to be better than in the national park. "The national park's policy of forbidding local consumptive use of resources and excluding local populations from making resource-related decisions, engendered animosity and considerable conflicts with the local populations. This created disincentives to local communities that are reflected in the condition of the forest. Decentralized decisionmaking, in this case, appears to be associated with better forest conservation outcomes" (Mwangi, Ongugo, and Njuguna 2000, 1).

Property rights create the incentive for people to invest in assets and give people possessions against which to borrow so that they might become entrepreneurs. Failure of a country's legal system to protect property rights will undermine the operation of a market exchange system. If individuals and businesses lack confidence that contracts will be enforced and the fruits of their labor protected, the drive to engage in productive activity declines along with the motivation to protect the environment. "Property rights makes the market possible," said Hernando de Soto, commenting on the establishment of property rights in post-totalitarian Iraq. "Once it's established, the world of credit comes along. It makes investment possible. Because when people invest, they are giving money for a property right. I imagine that the Iraqis are in the same situation of Egypt. Investment is not possible and credit is not possible.... So no property rights, no modern Iraq" (Ponnuru 2003).

Furthermore, people need access to economic opportunities. Access to capital and credit, for example, provided under conditions that promote economic opportunities creates an avenue for true development. Private savings and loans and government loan programs for farmers and students in the United States presented economic opportunities for low-income and working people. In South Asia, the Grameen banks have made capital available to poor people. By creating long-term, low-interest loans, the banks helped generate the wealth people needed to stimulate economic growth (Desta 1999).

Moreover, the institution of private property offers people an incentive to develop new technologies because individuals know they will benefit from investments they make in research and development. Tremendous energy and resources are being applied today to the development of practical substitutes for fossil fuels because the motivation exists to discover lower cost substitutes. For example, brokers are offering cash to farmers who are willing to plant a crop of wind turbines, and farmers are discovering that investing in wind power can be more profitable than raising traditional crops. Large companies are eager to harness the wind. Shell Oil, for example, recently bought its first wind farm in Wyoming. Landowners are also eager to collect wind royalties—especially those who can continue to farm with turbines on their property. As Pat Wood, President Bush's appointee to the Federal Energy Regulatory Commission, observed, "We've got lots of wind and it's about time that people figured out a way to make some money off it" (Huggins 2001, 45).

Economist Julian Simon continually drove home the point that human ingenuity is perpetually responding to impending scarcity by developing alternative technologies that mitigate against that scarcity. The key to mitigating natural resource constraints is to switch on human ingenuity, which allows us to accomplish more with a given amount of resources. The fall of the Berlin Wall and communism has made it clear that turning on this ingenuity requires getting the incentives right through the appropriate institutions. With property rights and the rule of law in place, economic growth and environmental improvements will follow (see Anderson 2003).

CONCLUSION

Institutional reform is not free, and many countries, for various reasons, resist reform that would improve problems related to human well-being. Perhaps the growing evidence that the protection of private property and growth-enhancing institutions are the building blocks of human well-being will persuade policymakers to reform their established systems (see Norton 2003). Only by upholding political and economic institutions that promote and protect property rights will we be able to sustain development and environmental quality. As the Friedmans put it:

> Our society is what we make it. We can shape our institutions. Physical and human characteristics limit the alternatives available to us. But none prevents us, if we will, from building a society that relies primarily on voluntary cooperation to organize both economic and other activity, a society that preserves and expands human freedom, that keeps government in its place, keeping it our servant and not letting it become our master. (Friedman and Friedman, 1980, 37)

It is critical that we focus our efforts on developing and protecting the institutions of freedom rather than on regulating human use of natural resources through political processes. The environment is getting better, not worse, and it will continue on this course if human ingenuity can continue to hammer out the institutions of freedom, namely property rights and the rule of law—institutions that will provide the incentive for us to solve whatever environmental problems might come our way.

As we head into the next millennium, it becomes increasingly clear that the progress we have enjoyed is primarily attributable to the freedom of the marketplace, and Milton and Rose Friedman have done much to ensure that we have come far on this path. It is important to continue their work by ensuring that the property rights path to sustainable development is made more visible in order to protect the institutions of freedom and the environment at the same time—only then can we have our environmental cake and eat it too!

REFERENCES

Anderson, Terry L., ed. 2003. *You Have to Admit It's Getting Better: From Economic Prosperity to Environmental Quality*. Stanford, CA: Hoover Institution Press.

Anderson, Terry L., and Laura E. Huggins. 2003. *Property Rights: A Practical Guide to Freedom and Prosperity*. Stanford, CA: Hoover Institution Press.

Bethell, Tom. 1998. *The Noblest Triumph: Property and Prosperity Through the Ages*. New York: St. Martin's Press.

Bhattarai, Madhusudan. 2000. "The Environmental Kuznets Curve for Deforestation in Latin America, Africa, and Asia: Macroeconomic and Institutional Perspectives." Dissertation, Clemson University, Clemson, SC, December.

Birch, Charles. 1976. "Creation, Technology and Human Survival: Called to Replenish the Earth." *The Ecumenical Review*, vol. XXVIII, January.

Bruntland, Gro Harlem. 1987. *Our Common Future: United Nations World Commission on Environment and Development*. Oxford, UK: Oxford University Press.

De Soto, Hernando. 2000. *The Mystery of Capital: Why Capitalism Triumphs in the West and Fails Everywhere Else*. New York: Basic Books.

Desta, Asayehgn. 1999. "Environmentally Sustainable Economic Development." *Journal of Political Ecology* 6: 236.

Ehrlich, Paul R., and Ann H. Ehrlich. 1996. *Betrayal of Science and Reason: How Anti-Environmental Rhetoric Threatens Our Future*. Washington, DC: Island Press.

Epstein, Richard. 1995. *Simple Rules for a Complex World.* Cambridge, MA: Harvard University Press.

Friedman, Milton. 2002. "Preface: Economic Freedom behind the Scenes." In *Economic Freedom of the World: 2002 Annual Report,* by James Gwartney and Robert Lawson, with Chris Edwards, Walter Park, Veronique de Rugy, and Smita Wagh. Vancouver, BC: The Fraser Institute.

Friedman, Milton, and Rose D. Friedman. 1980. *Free to Choose: A Personal Statement.* Orlando, FL: Harcourt.

Goklany, Indur. 2003. "Economic Growth, Technological Change and Human Well-Being." In *You Have to Admit It's Getting Better: From Economic Prosperity to Environmental Quality,* ed. Terry L. Anderson. Stanford, CA: Hoover Institution Press.

Gwartney, James, and Robert Lawson, with Walter Park and Charles Skipton. 2001. *Economic Freedom of the World: 2001 Annual Report.* Vancouver, BC: The Fraser Institute.

Gwartney, James, and Robert Lawson, with Neil Emerick. 2003. *Economic Freedom of the World: 2003 Annual Report.* Vancouver, BC: The Fraser Institute.

Hardin, Garrett. 1968. "The Tragedy of the Commons." *Science* 162:1243–48.

Hayward, Steven F. 2002. "Making Sense of Sustainable Development." *Brief Analysis No. 422.* Dallas: National Center for Policy Analysis.

Huggins, Laura E. 2001. "Wind for Sale." *Weekly Standard,* October 8.

International Institute for Sustainable Development. 2002. *The Sustainable Development Timeline,* 3rd ed. www.iisd.org.

Johannesburg Declaration on Sustainable Development: From Our Origins to the Future, September 4, 2002.

Lomborg, Bjørn. 2001. *The Skeptical Environmentalist.* New York: Cambridge University Press.

Meiners, Roger E., and Andrew P. Morriss, 2001. "Pesticides and Property Rights." *PERC Policy Series* Issue No. PS-22, May. Bozeman, MT: PERC.

Mwangi, Esther, Paul Ongugo, and Jane Njuguna. 2000. "Decentralizing Institutions for Forest Conservation in Kenya: Comparative Analysis of Resource Conservation Outcomes Under National Park and Forest Reserve Regimes in the Mt. Elgon Forest Ecosystem." Presented at "Constituting the Commons: Crafting Sustainable Commons in the New Millennium," the Eighth Conference of the International Association for the Study of Common Property, Bloomington, IN, May 31–June 4.

Norton, Seth. 2003. "Population Growth, Economic Freedom, and the Rule of Law." In *You Have to Admit It's Getting Better: From Economic Prosperity to Environmental Quality,* ed. Terry L. Anderson. Stanford, CA: Hoover Institution Press.

O'Riordan, Timothy. 1988. "The Politics of Sustainability." In *Sustainable Environmental Management: Principles and Practice*, ed. R. K. Turner. London: Bellhaven Press.

Panayotou, Theodore. 1997. "Demystifying the Environmental Kuznets Curve: Turning a Black Box into a Policy Tool." In *Environment and Development Economics* 2: 465–84.

Political Risk Services. 1997. *International Country Risk Guide*. East Syracuse, NY: Political Risk Services.

Ponnuru, Ramesh. 2003. "Who Should Own Iraq? An Interview with Hernando de Soto." *National Review*, May 5. www.nationalreview.com/ponnuru/ponnuru050503.asp.

Taylor, Jerry. 1993. "The Growing Abundance of Natural Resources." In *Market Liberalism: A Paradigm for the 21st Century*, ed. Ed Crane and David Boaz. Washington, DC: Cato Institute.

———. 2002. "Greenback in Jo-burg: The U.N.'s Latest Earth Summit." *National Review*, September 16, 29.

———. 2003. "Natural Resources." Online: www.sdnetwork.net/natural_resources.htm.

Tren, Richard. 2002. "The Stockholm Convention: Who Stands to Gain?" International Policy Network, Weekly Commentary, July 12. www.policynetwork.net/weekly_comment/richard_tren_july02.htm.

Töpfer, Klaus. 2002. Opening remarks at the sixth session of the Intergovernmental Negotiating Committee for the *Stockholm Convention of Persistent Organic Pollutants*. Geneva, Switzerland, June 17–21.

United Nations Development Program. 1997. *Human Development Report*. New York: Oxford University Press.

U.S. Department of Housing and Urban Development. 1995. *Community Sustainability: Agendas for Choice-Making & Action*. Washington, DC.

World Bank. 2001. *World Development Indicators*. CD-ROM. Washington, DC.

Yandle, Bruce, Maya Vijayaraghavan, and Madhusudan Bhattarai. 2002. *The Environmental Kuznets Curve: A Primer*. PERC Research Study RS-02-1. Bozeman, MT: PERC.

Economic Freedom and Environmental Quality

Richard L. Stroup

Milton and Rose Friedman conclude their 1980 book *Free to Choose* with a chapter entitled "The Tide Is Turning." It includes the optimistic statement that "we are waking up." Americans are "again recognizing the dangers of an overgoverned society, coming to understand that good objectives can be perverted by bad means, that reliance on the freedom of people to control their own lives with their own values is the surest way to achieve the full potential of a great society" (310).

This hopeful statement foreshadowed much of what has happened since then. A president, more appreciative of markets than any in decades, was elected in the United States that year, followed in Great Britain by the choice of a market-oriented prime minister. Ten years after *Free to Choose* was published, the Berlin Wall fell. In many ways the tide has indeed turned. The work of the Friedmans was one of the reasons.

Their impeccably reasoned arguments in favor of economic freedom, starting in a big and lasting way with *Capitalism and Freedom*, helped to gradually bring people in the United States and elsewhere to recognize the importance of economic freedom. In the United States, beginning in the late 1970s, trucking deregulation and the freeing of airline prices from regulation both brought sizeable, well-recognized benefits, as did the deregulation of oil prices. In some places, the move toward economic freedom went further. For example, Roger Douglas, finance minister in New Zealand's Labour government beginning in 1984, was able to cut income tax rates in half, deregulate wide sectors of the New Zealand economy, end farm and business subsidies, and privatize most state-owned enterprises there. The progress made in nations around the world was substantial, and in some respects—albeit in fits and starts—the trend continues. In England, the disasters of postwar socialism were, to a significant extent, reversed during the Thatcher era.

But today there is another, growing force at work in the opposite direction. Against the progress in the understanding of the importance of markets and economic freedom is running a worrisome tide: the growing impact of regulatory juggernauts stemming from environmental policy in the United States. Even as economic regulation in several cases declined, environmental regulation has increased.

ENVIRONMENTAL IMPACT

The role of the Friedmans and their books in promoting the public's understanding of property rights and markets specifically in the area of environmental policy is limited. They did begin their contributions early—with one of the first statements, perhaps, challenging the need for government support of a national park. In *Capitalism and Freedom* (1962, 31) they said about Yellowstone: "If the public wants this kind of an activity enough to pay for it, private enterprises will have every incentive to provide such parks." And they point out that unlike the case of city parks, to identify those who enjoy visiting them is not hard, nor is collecting revenue to support them. We at PERC—Terry Anderson and Don Leal in particular—and others have written extensively on how such a system can work and, as the Friedmans pointed out, has in fact been working.

But in 1962 environmental regulation was barely a blip on the radar screen of even most market-oriented economists. While the blip had grown larger by 1980, the year *Free to Choose* was published (and, by the way, the year PERC was founded), other concerns were still much greater for the Friedmans and most other economists. Yet Milton Friedman (no doubt with the help of Rose) made yet another contribution to the literature—a contribution that I believe is having a quiet but profound impact in helping us better recognize, demonstrate, and control the regulatory role of central governments, including—over time, I believe—the role of environmental regulation.

That contribution is the development of the Economic Freedom of the World (EFW) index. Milton Friedman's role in that project was, and is, a large one. Not yet fully recognized is its fundamental importance in helping us learn about the results of policy alternatives and to settle disagreements on the central government's proper role. One of these disagreements is the government's role in environmental policy.

For those of us interested in environmental policy options, the issue can be stated this way: Will environmental policy improve when experts from the central government control more of the nation's economy? Or will private property, protected in courts rather than by a central bureau and traded in markets, yield better environmental results? Put more appropriately, the question is this: *When* will each of these policy approaches work best? These questions on the environment are hotly debated today. But they are not unlike a set of questions about prosperity and economic growth that were debated throughout much of the twentieth century. Now, as then, good economic analysis focuses on the role of information and on incentives to find and use that information wisely.

We can learn from that long "socialist calculation" debate, and we now have a tool, the EFW index, that should help us find answers much more quickly than those to the previously unsettled questions. When economists who have a good grasp of how theory can help us understand and answer the real-world questions at hand, the index and its components, applied country by country along with other information, have great potential. They can be used to clarify and quantify the impact of "freedom to choose" and other policy options as they influence not only economic growth but other measures of human well-being, including environmental indicators—from health and longevity to the disappearance of species.

HISTORY AND ITS LESSONS

Beginning around 1920, a number of economists took part in what is now known as the socialist calculation debate over the productivity and, indeed, the feasibility of socialism. Ludwig von Mises and later F. A. Hayek were prominent in arguing that when governmental control replaced private property rights and markets, the quality of decisions would fall. Relative prices set in open markets would no longer be available to guide efficient production or even to identify the most appropriate goods to produce. Von Mises and Hayek questioned the ability of central planners to give rational guidance to the economy without the information generated and constantly updated by the price system that emerges from market trading of privately owned rights. Without true markets, how could planning really be rational?

It was not until decades later that many economists came to understand the importance of what von Mises and Hayek said. The centrally directed planning model seemed productive to many, perhaps most, economists until recently. For example, in 1985 the popular introductory economics textbook of Paul Samuelson and William Nordhaus put it this way: "The Soviet model had surely demonstrated that a command economy is capable of mobilizing resources for rapid growth and awesome military power." They did note that it had been done "in an atmosphere of great human sacrifice—even loss of life—and political repression." Whether the sacrifice was worth it, they said, was "one of the most profound dilemmas of human society" (Samuelson and Nordhaus 1985, 776). The basic lessons taught by von Mises, Hayek, and the Friedmans are slow to be absorbed, it would seem. Indeed, the teaching is tragically slow for people living under badly flawed systems—and for those of us living under governments that were importing some of those flaws.

After the fall of the Berlin Wall and the collapse of the Soviet Union (and with it the governments of many of its satellite nations), the same authors said much the same thing in the 1995 edition of their book, but the advantage of markets was now recognized. Samuelson and Nordhaus wrote that "it appears that in the modern world of open borders and high-quality manufactured goods, the blunt control of the command economy could not match the finely tuned incentives and innovation of a market economy" (716). The "finely tuned incentives," of course, come from the price system. When costly but successful innovation brings personal rewards to those who make it happen, more innovation is encouraged. When higher quality products earn a higher price, higher quality goods become more available. These signals and incentives are systematically missing from the socialist system.

With the fall of the Iron Curtain, observers could see the devastation left by the central planning systems. Markets came to be more appreciated and thus more utilized in the production of goods and services in much of the world.

Throughout much of the twentieth century, however, market socialists were viewed as winners of the intellectual debates, and the tide of history seemed to be on their side. According to Robert Heilbroner, who favored the socialist viewpoint, the debate was seemingly settled in 1940 by Oskar Lange (whom Heilbroner calls a "brilliant young economist"). Lange contended that a central planning board could solve the problem of economic calculation by keeping an eye on inventories and changing prices in response to changes in inventories (Heilbroner 1990).

The bleak outlook for capitalism led to the formation in 1947 of the Mont Pelerin Society, with Hayek as the founding president. Milton Friedman was a founding member and served as president from 1970 to 1972. The goal of the society, according to its web page (see www.montpelerin.org), was to "facilitate an exchange of ideas between like-minded scholars in the hope of strengthening the principles and practice of a free society and to study the workings, virtues, and defects of market-oriented economic systems." The society's "statement of aims" laid out their urgent concerns:

> Over large stretches of the earth's surface the essential conditions of human dignity and freedom have already disappeared. In others they are under constant menace from the development of current tendencies of policy. The position of the individual and the voluntary group are progressively undermined by extensions of arbitrary power. (www.montpelerin.org)

Members of the Mont Pelerin Society set about trying to put their concerns into action. An important manifestation of this goal was the publication in 1962 of *Capitalism and Freedom*, an extraordinary book for the time. It stirred interest and built support for the idea that economic decisions should be made by individuals acting on their own initiative, not forced to follow the dictates of governments. The major effects of *Capitalism and Freedom* were probably primarily on young people at the time and future leaders—including Ronald Reagan. However, the impact did not become fully visible until many years later. The logic was sound and convincing to many readers, but despite Milton Friedman's strong background in statistics, clear cross-country comparisons of the sort he was later to help make feasible were not yet available to make more obvious and more concrete the value of the points made in the book.

Decades later, after the fall of the Berlin Wall, however, the substance of what von Mises, Hayek, and the Friedmans had written became more obvious. By 1993, Heilbroner could write:

> Socialism—defined as a centrally planned economy in which the government controls all means of production—was the tragic failure of the twentieth century. Born of a commitment to remedy the economic and moral defects of capitalism, it has far surpassed capitalism in both economic malfunction and moral cruelty. Yet the idea and the ideal of socialism linger on.

Later in the same article he recognizes the source of the problem and what is needed to solve it:

> The main obstacle to real perestroika is the impossibility of creating a working market system without a firm basis of private ownership, and it is clear that the creation of such a basis encounters the opposition of the former state bureaucracy and the hostility of ordinary people who have long been trained to be suspicious of the pursuit of wealth.

The basic lesson of von Mises, Hayek, and the Friedmans had been learned by a formerly dedicated supporter of socialism. But without the availability of statistical tests and demonstrations using international data of the kind made available on a systematic basis today by the EFW index, the lessons had taken decades to be widely absorbed.

ECONOMIC FREEDOM OF THE WORLD INDEX

In the mid-1980s, Milton Friedman and Michael Walker, executive director of the Fraser Institute in Vancouver, Canada, began a project to help explain the various aspects of economic freedom. Most of all, the goal was to figure out ways to measure economic freedom and to determine the consistency of each government in providing or allowing them. Once the measures were identified, it would be possible to estimate the effects of the policies measured. Supported by the Liberty Fund, Friedman and Walker convened a series of six meetings of economists from 1986 to 1994. The goal was to devise ways to measure the economic freedom that was of such great concern to them, to the Mont Pelerin Society, and ultimately to all citizens—whether they recognized it or not. Many distinguished economists participated, including Nobel laureates Gary Becker and Douglass North.

The key product at the end of the series of meetings was the Economic Freedom of the World index. The first version of the index was published in 1996 by James Gwartney, Robert Lawson, and Walter Block. With the help of classical liberal institutes worldwide, they are seeking more and better data. Aided by continued guidance from Michael Walker and Milton Friedman, Gwartney and Lawson regularly update, improve, and extend the index.

The EFW index ranks economic freedom in 123 nations on the basis of objective, published, and available data. These data are selected to determine the extent to which (in the words of the latest EFW report) each country has institutions and policies that "provide an infrastructure for voluntary exchange" and "protect individuals and their property from aggressors seeking to use violence, coercion, and fraud to seize things that do not belong to them" (Gwartney and Lawson 2003, 5). Those in a country who seize things that do not belong to them may—and often do—include the government. Some of the criteria for economic freedom, of course, involve restraining the powers of those in government. The EFW index includes criteria data in five areas: the size of government, the legal structure and the security of property rights, access to sound money, freedom to exchange with foreigners, and regulation of credit, labor, and business.

When a nation's EFW index number is high, the market is playing a larger role relative to political control of the economy. More decisions are being made privately, coordinated in markets with less interference from government. This does not mean, however, that the government's role is less important; it is merely less extensive in scope, exerting little direct control over economic decisions. An essential role for government in a market economy is its protective function: the protection of persons and their property from theft, fraud, and violence.

Where the legal structure and security of property rights are stronger, government is doing a crucial job well, and the EFW index reflects this in a higher measured degree of economic freedom. Thus, the EFW index is valuable in research to help settle the arguments of the sort that propelled the socialist calculation debate.

The EFW index enables researchers to examine, and to demonstrate convincingly, how the economic freedom of a country affects that country's prosperity, growth, and poverty. A capsule view of the impact of economic freedom on these variables is given in the 2003 version of the index:

> Economic freedom is highly correlated with per-capita income, economic growth, and life expectancy. Increased economic freedom does not lead to greater income inequality. The lowest 10% of income earners in nations in

the bottom quintile of economic freedom receive 2.27% of total income in their nations; in nations in the fourth quintile, the bottom 10% receive 2.66% of total income; in the third quintile, 2.25%; in the second quintile, 2.83%; and in the top quintile, 2.68%. The actual income of poor people increases as nations gain in economic freedom because of the increased wealth economic freedom generates. The average per-capita income of the poorest 10% of people in nations in the bottom quintile is US$873 compared to US$6,681 for those in the top quintile.

This statement from the authors reflects the results of more than 135 papers (including several by Gwartney and Lawson), published in refereed journals, that use the EFW index or its components to explain various outcomes in the real world. The web site of the project (www.freetheworld.com) lists these publications along with several working papers and links directly to many of them. These articles and book chapters cover the impact of economic freedom on an even broader range of variables, from income to intellectual property and public health and the environment, typically accounting also for many other influences, sometimes including political freedom.

ECONOMIC FREEDOM AND THE ENVIRONMENT

Today, the importance of markets for prosperity and growth is being more widely recognized, but the role of markets in the environment is still often neglected. Economics principles texts most often discuss environmental problems as "market failure." The problem, of course, is that markets perform their function only when property rights are well-defined, enforced, and tradable. When the property rights of individuals—their rights against anyone who would violate their rights by theft, fraud, violence, or pollution—are not properly defined or defended, the fault does not lie with the market (which in this case is non-existent). It may, instead, lie with the government that failed to protect citizens against rights violations by others in society.

Environmental harms occur when there is no protection for individuals and their property against damage, including environmental damage. Yet the "market failure" explanation for environmental problems is common, and much environmental policy in the United States and around the world today is destructive both of property rights and of the market approach. Control by the government, especially the central government, is more and more the policy approach adopted.

Just as the economists of the twentieth century debated the question of collective vs. private control, a key debate today is over the question of whether increased governmental control helps or harms the environment in which we live. Fortunately, we have more tools and more experience today to help us settle the question, without relying entirely on trial and error, which took many decades in the case of socialism in the last century.

One way to research options for answering this question is to examine the effects of decreased economic freedom (that is, more government control) on the condition of our environment. A large number of articles and books with case studies have been published in the past, but more recently, data series such as the EFW index offer insight.

So far, only a few studies have been done on environmental questions using the EFW index. It is an area ripe for additional study. But studies have examined the effects of economic freedom, or some of its components, on some environmental measures in some groups of nations. These are statistical studies using economic freedom and an independent or explanatory variable. For example, in a chapter in *Who Owns the Environment?* economist Seth Norton (Norton 1998a) found that in nations where property rights (as measured by the EFW component) are strong, various measures of environmental quality (as measured by World Bank data) are higher than in nations in which property rights are weak.

Norton used three other measures of property rights that led to similar results and the same conclusion, and the results were statistically significant. Access to clean water, sanitation measures, life expectancy, and deforestation all are more favorable in nations with stronger private property rights. When property rights were well protected, for example, about 90 percent of the population had access to safe water; but in nations with weak property rights, only about 60 percent of the people had that key health advantage.

Norton (1998b) has also examined the impacts of property rights on the poorest people of the world. One measure of poverty he used was the United Nations' Human Poverty Index (HPI), which includes the environmentally related elements of longevity as well as access to safe water, among other measures of well-being. The HPI is a distinctive database that considers the conditions of only the most deprived people in a nation's communities. Using this database, Norton finds that the influence of stronger property rights is substantial and positive: "Where property rights are strong, the HPI is substantially reduced," he writes, and "weak rights are associated with greater deprivation for the world's impoverished" (239). For poor people in poor nations (as Norton had found across nations in general), stronger property rights, an important component of economic freedom, improve the environment, health, and other aspects of citizens' lives.

POLITICAL FREEDOM AND THE ENVIRONMENT

The EFW index and similar databases have allowed researchers to provide evidence that economic freedom leads to higher environmental quality. Even in modern democracies with largely free markets, there is evidence that reducing economic freedom in favor of political control of environmental decisionmaking can reduce environmental quality. Indeed, a number of case studies have shown that moving away from property rights protected by the common law toward statutory pollution policy with environmental regulations administered bureaucratically has allowed special interests to capture parts of the regulatory regime for their own advantage, sometimes to the detriment of the environment.

In 1981, Ackerman and Hassler showed in *Clean Coal, Dirty Air* that the Clean Air Act Amendments of 1977 were shaped by Eastern high-sulfur coal interests. These companies and miners' unions successfully pressured Congress to specify the use of scrubbers, which virtually required the use of high-sulfur coal in coal-fired electric power plants—even where lower-sulfur coal would have been cheaper and would have reduced sulfur emissions by greater amounts. Economic freedom was reduced in the name of air quality improvements, and in some cases the air was made dirtier as well. Even with scrubbers in place, high-sulfur coal could produce more sulfur oxides than a new plant burning clean coal would have done. The democratic process had changed environmental policy, but not for the better.

In his book *The Political Limits of Environmental Regulation* (1989), Bruce Yandle showed how pressure to replace common law with statute-based regulations came not primarily from victims of pollution but from special interest groups often seeking advantage over their competitors. Such regulations are brought about politically in a democracy, but they cannot be expected to be efficient or cost-effective in improving environmental quality.

Elizabeth Brubaker (1998) has revealed that in Canada, too, rent-seeking special interests used the democratic process to gain for themselves at the expense of the environment. In an earlier book (1995), she compiled impressive evidence that Canada's movement away from property rights and common law to government regulation under statute had, on balance, degraded that nation's air and water resources.

Michael Stroup, in a recent working paper, affirms the environmental benefits of economic freedom for industrialized countries, using data from the thirty Organisation for Economic Co-operation and Development (OECD) nations. His paper has an interesting twist that lends support to the idea that political control can hurt the environment on balance, not only in the case of poor and socialist nations, but also in a modern democracy. Stroup studied the impact of economic freedom on each OECD nation's emissions (measured per unit of economic activity) of four air pollutants: sulfur oxides, oxides of nitrogen, visible particulates, and carbon dioxide. Using multiple regression analysis to account for other factors, and studying four time periods from 1980 to 1995, he finds that these measures of environmental performance improve when economic freedom, as measured by the EFW index, is greater.

Stroup's analysis also includes political freedom, using measures indicating the ability of citizens democratically to influence the nation's policies, including environmental policies. He finds that for more than half of all OECD countries, more political freedom (that is, more democratic political influence) leads to more air emissions per unit of output, not less. Indeed, he points out that "a greater level of political freedom within an OECD country tends to decrease the level of all four types of air pollution per dollar of GDP only when the level of economic freedom in that country is relatively low" (23).

Worldwide, political freedom gives all citizens an influence over government, including policy on the environment. Usually, this is a force for a better environmental outcome because governmental leaders are being held politically accountable for their actions. But when economic freedom is high, then there is much to lose, and more political freedom can work to the advantage of special interests. If the environment already is relatively clean, then democracy leaves a good deal of room for mischief by rent-seekers. It appears both from case study data and from Stroup's results that while the effect of economic freedom on environmental quality is consistently positive, the positive environmental effect of political freedom is conditional on the absence of a high level of economic freedom. When elected officials are not constrained by strong constitutional limits, the democratic political system can be used to transfer rents at the cost of other goods, including environmental quality.

GOVERNMENTAL INVOLVEMENT

Although the relative roles of economic freedom and political freedom on environmental quality are beginning to be understood, the evidence is limited in quantity and coverage. At this point, few are knowledgeable and even fewer are persuaded that property rights and markets have strong advantages over the regulatory state in the case of the environment. The situation cries out for more complete and more thorough research, as evidenced by the fact that environmental policies continue to move away from reliance on property rights and economic freedom, toward regulatory decisions and control.

To conduct this research properly, it is necessary to have a more complete theoretical treatment of the problems than we see in most classrooms, where the "market failure" paradigm so often rules. The importance of property rights is becoming better known in the context of the environment, but public choice insights are seldom integrated into discussions of the environment, and the same is true of the information problem that was the focus of Austrian economists von Mises and Hayek. The better-informed theory is needed both to formulate testable hypotheses and to help researchers identify and assemble the necessary databases, just as the EFW process did over many years with the help of dozens of accomplished and experienced economists. This section seeks to explain parts of the puzzle: Why do regulatory policies, even when carried out by intelligent, hard-working public servants dedicated to their missions, turn out to be so costly and yet, too often, actually harm the environment? If property rights and markets are the basic policy in a nation like Canada or the United States, why might more involvement by democratic government lead, arguably, to worse results? Several factors can help explain this.

1. As in the problem of socialist calculation in the former Soviet Union, regulators face the problem of information that is missing due to the lack of market trading. They also lack incentives to find and utilize the information needed for effective, cost-efficient regulation, especially when finding it is difficult and using it is costly.

To begin with, resources are limited, so regulators must decide how to prioritize environmental problems. Identifying the worst chemical risks or determining which species should be protected first is complex and difficult. Opinions will differ, and there is no market in which people who feel strongly about one position or another can bid for what they want.

Once a priority is set, regulators must decide how best to reach the goal. But once again, there is no lineup of offers from competing suppliers—to clean up at a given cost, say, or to provide habitat for a given animal or suite of species. And because we cannot compare the cost of offers to supply against offers to pay for what is supplied, there is no way to identify a stopping point where further action toward that goal is too costly to be warranted by the benefits produced.

There is, however, a regulator who, like his counterpart in a socialistic system, orders people to do something—to clean up chemicals or provide habitat without payment. Of course, a regulatory order that simply stops a proven violation of someone else's rights is an appropriate order without payment. But unlike a complainant asking relief against a polluter under common law, the typical regulator normally faces no burden of proof in determining whether a person's rights have been violated. The regulator often operates with what Justice Stephen Breyer has called "tunnel vision" (Breyer 1993). The regulator sees clearly only the task at hand, not the costs imposed on others by a regulation. This regulator has little incentive to hold back on using regulatory authority even though more costly responses are required to produce still more safety (or more habitat). The cost is usually borne by the regulated party, so the regulator has an incentive to seek even small improvements with high costs. Excessive regulation can result. On the other hand, if a politically organized special interest demands that the regulator divert his attention to other issues, regulators may well go along. Why pay a high cost to fight back? Better, perhaps, from the regulator's viewpoint to seek other margins to reduce the risk being regulated, to avoid conflict with a politically important regulated party. In such cases, regulation may be too lax to protect the public from serious risks.

A well-documented case where tunnel vision leads to regulations tight enough to harm the environmental mission is the case of land-use regulations under the Endangered Species Act (ESA). As currently applied, the ESA can be quite costly to landowners, giving them negative incentives to protect species. The possibly draconian penalties that landowners will experience as a result of using their land while having endangered species on the property lead them to change their land management. They can usually find easy ways to modify their habitat to reduce the likelihood that the listed species will find it attractive and thus be present.

Landowners naturally prefer to maintain management authority. Under the current rules, the populations affected are likely to be seriously harmed by such preemptive habitat modification. Each landowner has reason to learn what a listed species in the area likes or needs, to tweak land management practices to make what that species likes largely unavailable, and to inform neighbors about these practices. A resident population of an endangered species can lead the Fish and Wildlife Service to impose land management controls under the ESA. Both anecdotal (Stroup 1995) and statistical evidence (Lueck and Michael 2003) support this conclusion. The penalties of the ESA give landowners an incentive to manage their land *against* the listed species.

2. In a private setting, Coasian bargaining reduces the costs of reaching objectives, but such opportunities are typically lacking in a regulatory setting.

Once a regulatory decision is made, there is typically no legitimate way to bargain around it. A regulatory decision that costs the regulated party $10,000 but produces just $1,000 worth of benefits to the regulator's mission is wasteful, but it is likely to stand because the regulator achieves the benefit and doesn't pay the cost. Contrast this with the private sector. After a disputed property right is adjudicated and the right is determined, that right will still tend to flow to the highest-valued user—even if the right was not awarded to the highest-valued user. If the polluter owes a duty to stop the polluting activity, but stopping costs $10,000 while accepting the pollution would only cost the receptor $1,000, then we can expect the polluter to buy permission to pollute from the receptor at a bargained price higher than $1,000 but lower than $10,000. Neither inefficient pollution nor inefficient control need occur, if polluter and receptor find trading to be mutually beneficial. And when values later change, then the retrading of rights can allow peaceful and efficient adjustment. But under statutory regulation, when such exchange is typically not allowed, even the most inefficient order must be followed. That follows in part from the fact that even citizens not materially affected by the pollution may be allowed to enter the case as "stakeholders." In this case, trading to reduce the cost of pollution plus the cost of control is not likely to be feasible because there are too many stakeholders for whom stopping the pollution has some value, however small, but their voices may have public policy impacts out of proportion to any damages they might suffer.

Because in a market ownership rights can be traded or retained at will, there is little incentive in a market for either a buyer or a seller to posture or adopt sanctimonious attitudes and condemn other user demands as frivolous, as so often happens in discussions over the use of politically controlled lands, such as federal lands in the United States. An experience of the National Audubon Society illustrates the contrast between the constructive nature of private negotiations and the contentious nature of political discussions.

Officials of the Audubon Society are outspoken and hostile in their arguments against oil drilling on a federal wildlife refuge in Alaska. Yet they have worked comfortably and peaceably with the private oil company that they have allowed to produce natural gas on the Paul J. Rainey Preserve, which the National Audubon Society owns in Louisiana (Snyder and Shaw 1995). Gas was extracted only after the producers met Audubon's strict stipulations. Audubon used the resulting revenue to enhance its mission on the refuge and elsewhere. Audubon has the right to determine what happens on its land, and it has strong incentives to avoid risking the loss of support from its members by allowing damage to the habitat it owns; but it also has the right to gain support for its mission by producing petroleum. Audubon's mission can be given a net gain by natural gas revenues that contribute more than the tiny losses to existing habitat resulting from the careful petroleum extraction procedures.

Without trade, results are less efficient. This reduced efficiency harms environmental quality and environmental policy in two ways. First, less efficiency reduces wealth, and when wealth declines, the willingness and ability of those affected to demand environmental quality decline. This income or wealth effect has been estimated by Donald Coursey to be 2.5 times as strong as the change in income causing the income effect. The estimated income elasticity of demand, that is, is 2.5. Second, an environmental policy that is less efficient has a price effect, too. The policy delivers less "bang for the buck," and voters will demand less of a policy when the cost of that policy's results costs them more (Coursey 1993).

3. Public decisions are public goods: Accountability in the public sector is largely missing as a result, and free riders are evident at every level.

Gordon Tullock made this point more than thirty years ago (Tullock 1971, Stroup 2000). The most fundamental reason for poor accountability in government and the presence of free riders is that voters are rationally ignorant. An individual, knowing that one person does not determine the outcome of an election, is likely to spend more time and attention deciding which car to buy— or even which tennis racket—than on which candidate to support. This is a rational choice for the individual, but it also means that voters are not able to hold government responsible in an informed way.

The impact on environmental policy can be seen through some interesting research about how people respond to risks. It is well known in the risk analysis community that members of the general public systematically underestimate common and significant risks but overestimate small environmental risks of the sort commonly regulated. But this bias disappears when the risks that each person is asked to estimate are the specific kind that person faces, as when an elderly person is asked to estimate the risk of death from a slip and fall, a common danger only for older people (Benjamin and Dougan 1997). People know much more about risks commonly faced by themselves, their families, or their friends. In contrast, as voters they affect government decisions about many matters on which they are largely ignorant. Thus, putting voters in general in charge of environmental risks guarantees that the risk management system—voters are ultimately in charge of the system—will be "flying blind" much of the time.

The "free rider" problem of public policy has many implications. With the public largely uninformed, special interests and "stakeholders" can use activist tactics and the resulting publicity to stop a policy they do not like. The Environmental Protection Agency, for example, has been shown to make decisions influenced by press coverage of its proposed rules (Yates and Stroup 2000). Similarly, federal regulators respond to media coverage when deciding about public lands. In contrast, if the stakeholders had some true ownership, so they could sell their interests and other stakeholders could not step in, then quite possibly the stakeholders could reach a mutually beneficial result. But instead, non-owners are allowed to usurp some of the rights of "owners," and almost anyone is considered a "stakeholder" and has the standing needed to bring on a de facto veto of use. In the case of the Alaskan wildlife reserve, the National Audubon Society gives up nothing when it helps to stop trade and prevent drilling and production by oil producers. Unlike the case of a preserve it owns, it does not give up money or other value when it acts to stop drilling on federal lands.

Decisions will be wiser when they are made privately by individuals who gain personally and substantially from their own resource conservation and pay the major cost personally when they waste resources. Where regulation is the only realistic option, however—think of auto air pollution in the Los Angeles basin—devolution of regulation to the lowest possible level can concentrate both benefits and costs closer to where the decisionmakers live and the knowledge base of the relevant citizens is better, enabling them to hold their local government more accountable. Relying on common law—using the courts to protect individual rights—will also lead to better information, when reliance is feasible. Because common law demands a burden of proof and follows standards of evidence, information that will stand up to cross-examination is necessary to bring the force of the law to bear. In contrast, the publicity campaigns that affect governmental regulatory decisions have no such burdens of proof. Instead, cheap talk may rule. The environment is likely to be one of the victims in such a case.

4. A good produced in the private sector is likely to be better, as judged by its users, than the same good produced in the public sector.

The Friedmans were right when, as noted above, they wrote about Yellowstone: "If the public wants this kind of an activity enough to pay for it, private enterprises will have every incentive to provide such parks." Not only do private enterprises have an incentive to provide goods and services such as parks, but private provision tends to provide greater benefits as well. That is because those who pay—and in the private sector, visitors usually pay the full cost of the services they receive—will control. Where customers pay, we can expect the goods and services to be better targeted to those who want them most and to be provided more cost-effectively. Evidence from state parks, where customers pay a much larger portion of the total costs than in national parks, supports this expectation.

Research at PERC by Donald Leal and Holly Fretwell has shown that fiscal difficulties have been causing both national and state parks to move toward more reliance on revenues received from user fees and from ancillary suppliers to users—such as concessionaires. "New Hampshire and Vermont state parks are already self-supporting, and a growing number of others are headed in that direction. An entrepreneurial spirit has taken hold in Texas, South Dakota, and Arkansas. Park managers have developed a myriad of new programs, activities and events for which they charge affordable fees. The response has been positive. Visitation has increased, and so have revenues" (PERC 2003). When park managers derive their support from visitors and other voluntary supporters, they are motivated to provide those supporters with good services and products at a low cost.

A large natural experiment was conducted two centuries ago that is relevant to this discussion of private versus public provision, even though it was not directly related to environmental policies. Economist Kelly Olds, in a 1994 *Journal of Political Economy* article, discussed the impact of "disestablishment" of state churches in the United States. Around 1800, one state at a time, the young nation turned away from state churches. All tax support of churches was ended. A surprising thing happened—surprising to many of us, anyway: Church attendance, church budgets, and the number of preachers did not shrink; instead, all grew substantially. (Olds examined Connecticut and Massachusetts in detail.)

To this day, America is one of the few industrial democracies without a state-supported church. America is also far and away the *leader* in church attendance in this group of nations and the *leader* in religiosity. When society turned to the market order rather than government support, preachers, church leaders, and small groups of those most concerned and most faithful swung into action. Religion lost government support but gained far more. There is no mystery about why. A private church or a club is not run by the average voter or by the deliberations of a legislature. It is run by those who care most about the church or club and its mission. As a result, churches are more diverse and better operated, and the survivors thrive.

CONCLUSION

The four points made above on government involvement should help us to see the pitfalls in turning more and more authority over to centralized government as nations become richer and demand ever-increasing environmental quality. While regulation often seems to be a way to obtain what we want at low cost, these points suggest that the actual results of new regulations may run counter to their stated intentions, as happened with socialism in the twentieth century and as happens today with environmental regulation. Accountability for costs and rewards for benefits generated are hard to achieve in government.

We at PERC are dedicated to the belief that while policy is seldom made by economists, nor made just as they would recommend, economists and their ideas do have serious consequences over time. It is worth doing the economic research, doing it well, and doing it extensively in policy-relevant areas. The EFW index and the research based on it are models for what is needed in the new intellectual wars over the socialist model of centralized control now being applied in the name of the environment.

Today, we must ask whether the tide is turning back from the progress brought on after many decades of intellectual battle by von Mises, Hayek, and the Friedmans. After decades of suffering by millions of people—suffering noticed by most of the world only after the fall of the Berlin Wall and the failure of socialist nations—nations became more free; but will that freedom continue to grow?

On the bright side, there is evidence of continuing progress. The EFW 2003 annual report states: "Economic freedom continues to gain ground around the world" (Gwartney and Lawson 2003). Lessons learned from the painful decades of abuse heaped on citizens by socialist leaders, plus the knowledge from research based on the EFW index and similar indices, are having a real and continuing effect.

One place where the picture is not so bright, however, and where the tide is probably running the other way, is environmental policy. The claim is made that market failure is at the root of environmental problems and that market replacement by tighter governmental controls is the best solution.

These claims must be answered. Even at these early stages, the results from research based on the EFW index and its components, and on similar indices, are heartening. They verify what Milton and Rose Friedman, as well as other classical liberals including the researchers at PERC, have been saying about the usefulness of private property rights and the markets. Environmental protection and conservation depend upon the incentives provided by private property rights and the exchange of rights through markets. To spread this message widely, much more work must be done, of just the sort done so well by the Friedmans over the past several decades.

REFERENCES

Ackerman, Bruce A., and William T. Hassler. 1981. *Clean Coal, Dirty Air*. New Haven: Yale University Press.

Bernstam, Mikhail. 1991. *The Wealth of Nations and the Environment*. London: Institute of Economic Affairs.

Breyer, Stephen. 1993. *Breaking the Vicious Circle*. Cambridge: Harvard University Press.

Brubaker, Elizabeth. 1995. *Property Rights in Defence of Nature*. Toronto: Earthscan Publications.

———. 1998. "The Common Law and the Environment: The Canadian Experience." In *Who Owns the Environment?* ed. Peter J. Hill and Roger E. Meiners. Lanham, MD: Rowman & Littlefield.

Coursey, Donald. 1993. "Demand for Environmental Quality." Paper presented at the annual meeting of the American Economic Association in Anaheim, California, January.

Friedman, Milton (with the assistance of Rose D. Friedman). 1962. *Capitalism and Freedom*. Chicago: University of Chicago Press.

Friedman, Milton, and Rose D. Friedman. 1980. *Free to Choose: A Personal Statement*. New York: Harcourt Brace Jovanovich.

Gwartney, James, and Robert Lawson, with Neil Emerick. 2003. *Economic Freedom of the World: 2003 Annual Report*. Vancouver, BC: The Fraser Institute.

Heilbroner, Robert. 1990. "After Communism." *The New Yorker*, Sept. 10, 92.

———. 2002 (reprinted from 1993 version). "Socialism." In *The Concise Encyclopedia of Economics*, ed. David R. Henderson. www.econlib.org/library/Enc/Socialism.html.

Lueck, D., and J. A. Michael. 2003. "Preemptive Habitat Destruction under the Endangered Species Act." *Journal of Law & Economics* 46 (1): 27–60.

Norton, Seth W. 1998a. "Property Rights, the Environment, and Economic Well-Being." In *Who Owns the Environment?* ed. Peter J. Hill and Roger E. Meiners. Lanham, MD: Rowman & Littlefield, 37–54.

———. 1998b. "Poverty, Property Rights, and Human Well-Being: A Cross-National Study." *Cato Journal* 18 (Fall): 233–45.

Olds, Kelly. 1994. "Privatizing the Church: Disestablishment in Connecticut and Massachusetts." *Journal of Political Economy* 102 (April): 277–97.

PERC. 2003. "National Parks." *PERC Issues*. Downloaded Oct. 8 at: http://perc.org/publications/issuesinbrief/issue_natparks.php?s'2.

Samuelson, Paul A., and William D. Nordhaus. 1985. *Economics*, 12th ed. New York: McGraw-Hill.

———. 1995. *Economics*, 15th ed. New York: McGraw-Hill.

Snyder, Pamela, and Jane S. Shaw. 1995. "PC Oil Drilling in a Wildlife Refuge." *Wall Street Journal*, Sept. 7.

Stroup, Michael D. 2003. "Separating the Influence of Economic and Political Freedoms on Air Pollution: An Empirical Analysis of OECD Countries." Unpublished paper presented at the annual meeting of the Association of Private Enterprise Education, Las Vegas, April.

Stroup, Richard L. 1995. "The Endangered Species Act: Making Innocent Species the Enemy." PERC Policy Series PS-3, Bozeman, MT.

———. 2000. "Free Riders and Collective Action Revisited." *The Independent Review* 4 (Spring): 485–500.

Tullock, Gordon. 1971. "Public Decisions as Public Goods." *Journal of Political Economy* 79 (July–August): 913–18.

Yandle, Bruce. 1989. *The Political Limits of Environmental Regulation*. Westport, CT: Quorum Books.

Yates, Andrew J., and Richard L. Stroup. 2000. "Media Coverage and EPA Pesticide Decisions." *Public Choice* 102: 297–312.

Session 3

The Economic Burden of Taxation
William A. Niskanen

*The Transition to Private Market Provision
of Elderly Entitlements*
Liqun Liu, Andrew J. Rettenmaier, and Thomas R. Saving

The Economic Burden of Taxation

William A. Niskanen

The many burdens of government include those attributable to taxation, monetary policy, regulations, and restrictions on civil liberties. This paper is specific to the economic burden of taxation, without in any way minimizing the other types of burdens.[1]

THE MODEL

The economic burden of taxation is a function of three conditions: the level of the average tax rate, the relation of the marginal tax rate to the average tax rate, and the response of the tax base to changes in the marginal tax rate.[2]

Start with the basic relation of the size of the economy to the two major fiscal decisions:

(1) $$Y = aG^b(1 - R)^c$$

where Y = GDP per potential worker,
 G = expenditures for government services (excluding defense) per potential worker, and
 R = the *average* tax rate.

The two major fiscal decisions, of course, are the level of expenditures for government services (excluding defense) and the level of the average tax rate. This equation is expressed in terms of output per potential worker to capture the effects of G and (1 − R) on both hours worked per potential worker and on output per worker hour. The implicit assumption in this equation is that government expenditures for defense, transfer payments, interest payments, and subsidies have no significant net effect on the output per potential worker; there is ample evidence, of course, that most transfer payments reduce output per potential worker. The elasticity *c*, as I will demonstrate, reflects the combined effects of the relation of the marginal tax rate to the average tax rate times the elasticity of Y with respect to one minus the marginal tax rate. For the moment, pay no attention to the G variable; at the end of the paper, I will return to discuss the effects of the combination of G and R on the optimal size of government.

Given Equation 1, tax revenues per potential worker are

(2) $$T = RY = aG^b R(1 - R)^c$$

and the output per potential worker net of taxes is

(3) $$N = Y - T = aG^b(1 - R)^{(1+c)}.$$

The marginal economic burden of taxation, thus, is the change in net output per unit increase in tax revenues. Some manipulation of Equations 1 and 2 yields the following equation for the marginal economic burden:

(4) $$\partial N/\partial T = -[(1 + c)(1 - R)]/[1 - (1 + c)R].$$

Equation 4, plus the observed data for the average tax rate and an estimate of the elasticity c, thus, is sufficient to estimate the quantitative magnitude of the marginal economic burden of taxation. If the elasticity $c = 0$, of course, the reduction in net GDP is equal to the increase in tax revenues, and there is no deadweight loss of the additional taxes. The marginal cost of taxation, however, increases rapidly as a function of both c and R.

Before presenting my estimates of the relevant parameters, however, I promised to address the effects of the structure of the tax system, more specifically the relation between the marginal and average tax rates. A more precise formulation of Equation 1 would include the marginal tax rate M rather than the average tax rate R in the term in parentheses. For two reasons, however, I have chosen to use the average tax rate R: There are no available data on the income-weighted aggregate marginal tax rate or agreed procedures for estimating this rate. And, since T = RY, the average tax rate must be used in Equation 2, adding an undetermined variable to the model. If the marginal tax rate M, however, is a function of the average tax rate R, the average tax rate can be used in both equations. For example, if

(5) $$M = -x + yR,$$

then

(6) $$(1 - R)^c = (1 + x - yR)^z$$

and the relation between c and z is

(7) $$c = yz[(1 - R)/(1 + x - yR)].$$

The elasticity c, thus, is seen to be the product of the marginal effect of R on M and the marginal effect of $(1 - M)$ on Y. An increase in the elasticity c may reflect either an increase in the progressivity of the tax structure (the parameter y) or an increase in the adverse economic effect of the marginal tax rate (the parameter z).

THE ELASTICITIES

For this study, the elasticities b and c of Equation 1 are estimated by two independent techniques. The first technique is to estimate the long-term relation between fiscal choices and economic outcomes in the United States. For this purpose, several economic growth equations were estimated by two-stage least squares regressions, based on a sample of annual U.S. data from 1964 through 1999. Equations were estimated for the annual change of three dimensions of economic growth:

- real GDP per potential worker,
- output per hour in the business sector, and
- hours worked in the business sector per potential worker.

The primary fiscal measures in these regressions are the annual change in real expenditures for government services (excluding defense) per potential worker, one minus the average tax rate, and one minus the average tax rate in the second prior year. The first equation is sufficient to estimate the elasticities for this study, but the other equations were estimated to identify the relative effects of the fiscal choices on productivity and hours worked.

The second technique is to estimate the elasticities b and c that are implicit in the actual levels of G and R for the United States in 1996, given the model of the fiscal choices of democratic governments developed in my book. In effect, this involves solving my model of democratic government backwards from the known fiscal choices in 1996 to the elasticities that are consistent with these choices. The finding that the elasticities estimated by these two techniques are quite close may be an indirect validation of my model of the fiscal choices of democratic governments.

Table 1 presents the estimates of the elasticities of fiscal effects from these two techniques.

The other regressions on the changes in productivity and hours worked suggest that about half of the effect of tax changes on short-run economic growth operate through changes in productivity and about half through changes in hours worked. The estimated effect of the after-tax rate on hours worked is consistent with a large number of other studies, most of which have neglected to estimate the effect on productivity. In the long run, however, about two-thirds of the effect of tax changes on economic growth operate through changes in productivity, because there is no significant difference between the short-run and long-run effects on hours worked. The finding that the implicit estimate of the elasticity c is quite close to the short-run estimate from the time-series regression may suggest that our government takes into account only the short-run effects on output of changes in taxes.

THE MARGINAL ECONOMIC BURDEN OF TAXATION

Now we can address the primary topic of this paper. Table 2 presents estimates of the marginal economic burden of taxation from Equation 4 for a range of the variable R and the elasticity c. These numbers, again, are the marginal reduction in output (or income) after taxes per additional dollar of government tax revenue.

Given that the elasticity c implicit in recent U.S. fiscal conditions is about 0.8 and the average tax rate is about 0.3, the marginal cost of government spending and taxes in the United States may be about $2.75 per additional dollar of tax revenue. One wonders whether there are *any* government programs for which the marginal value is that high. Given the estimate of the long-term elasticity c from the U.S. time-series data, the marginal cost of government spending and taxes may be as high as $4.50 at the current average tax rate. The cost estimate in a benefit-cost study of any program financed by general taxes

Table 1
Estimates of the Elasticities of the Fiscal Effects on Economic Growth

	Estimated	**Implicit**
b	.200	.220
	(.036)	
c (short run)	.748	.772
	(.127)	
c (long run)	1.212	
	(.164)	

NOTE: Numbers in parentheses are the standard errors of the estimates from the regression equation.

Table 2
The Marginal Economic Burden of Taxation

		R	
	.2	.3	.4
c			
.4	1.556	1.690	1.909
.8	2.250	2.739	3.857
1.2	3.143	4.529	11.000

should be multiplied by the relevant number from this table. All of these estimates, of course, increase as a function of both c and R and approach infinity as R nears the revenue-maximizing tax rate.

SOME OTHER INTERESTING ESTIMATES

The model and the empirical estimates of Equation 1 also provide a basis for estimating several other interesting magnitudes: the revenue-maximizing average tax rate, the net output maximizing level of G, and the net excess burden of maintaining a 30 percent average tax rate given the optimal level of G.

The maximum average tax rate is determined from Equation 2 by setting the derivative of T with respect to R equal to zero; this yields the following equation for the maximum R:

(8) $$R = 1/(1 + c).$$

As the equation indicates, the revenue-maximizing average tax rate declines with an increase in the elasticity c, whether caused by an increase in the progressivity of the tax system or an increase of the elasticity of output with respect to one minus the marginal tax rate.

The level of G that maximizes net output is determined from Equation 3 by setting the derivative of N with respect to G equal to zero; this yields the following equation for the optimal ratio of G to Y:

(9) $$G/Y = b/(1 + c).$$

As this equation indicates, the optimal domestic spending share of GDP increases with the elasticity b and declines with the elasticity c. This has always presented somewhat of a dilemma for tax reformers; a reduction of the progressivity of the tax structure is likely to lead to an increase in the relative size of government spending because it reduces the marginal cost of additional spending. My own suggestion is that approval of any broad-based, flat-rate tax reform should be accompanied by a change of the voting rule to require a supermajority vote for any subsequent increase in the base or rate.

The net excess burden of taxation is also estimated from Equation 3 by calculating the net output if R = .3 (roughly what it has been in the United States for some years) relative to the net output if R is sufficient to finance only the optimal level of G. This is a rough estimate of the loss of net output from setting an average tax rate sufficient to finance government spending for defense, transfer payments, etc., in addition to the optimal level of G.

Table 3 presents these other interesting estimates for several levels of the elasticity c. All of the calculations of the optimal level of G/Y and the net excess burden are based on the elasticity $b = .2$, as there seems little uncertainty about this elasticity.

Table 3
Some Other Interesting Estimates

c	Maximum R	Optimal G/Y	Net Burden
.4	.714	.143	.247
.8	.556	.111	.350
1.2	.455	.091	.437

As expected, the revenue-maximizing average tax rate declines sharply with an increase in the elasticity c; this rate would be the peak of any Laffer curve expressed in terms of the average tax rate and is the ultimate limit on the sustainable level of government spending relative to GDP. The optimal level of expenditures for government services (excluding defense) relative to GDP also declines with an increase in the elasticity c but to a level that is not much lower than recent experience; in 2001, for example, government consumption expenditures and gross investment, excluding defense, were 14.5 percent of GDP. Given an estimate of the elasticity c that reflects the effects of the after-tax rate on both the supply of labor and on productivity, the optimal level of G is about 10 percent of GDP, a relative level of G that Milton Friedman has supported for many years. The net excess burden of taxation beyond that necessary to finance the optimal level of G, however, increases with the elasticity c. This column indicates that the net economic cost to the economy of a level of total spending and taxes beyond that necessary to finance the optimal level of G increases from about 25 percent of net potential output if $c = .4$ to about 44 percent of net potential output if $c = 1.2$. This does not suggest that there is no value to government spending above the optimal level of G, only that the net cost to the economy of this spending is much higher than the direct expenditures for these programs.

For those of you who may wish to pursue these issues in the larger context of the fiscal choices of alternative political regimes, I encourage you to read my book.

NOTES

[1] Most of this paper is a summary of some footnotes in my new book, *Autocratic, Democratic, and Optimal Government: Fiscal Choices and Economic Outcomes*, published by Edward Elgar in February 2004.

[2] The standard reference article on this issue is by Edgar K. Browning (1987), "The Marginal Welfare Cost of Taxation," *American Economic Review* 77: 11–23.

The Transition to Private Market Provision of Elderly Entitlements

Liqun Liu, Andrew J. Rettenmaier, and Thomas R. Saving

> *Workers paying taxes today can derive no assurance from trust funds that they will receive benefits when they retire. Any assurance derives solely from the willingness of future taxpayers to impose taxes on themselves to pay for benefits that present taxpayers are promising themselves. This one-sided 'compact between the generations,' foisted on generations that cannot give their consent, is a very different thing from a 'trust fund.' It is more like a chain letter.*
> —Milton and Rose Friedman, *Free to Choose*, 104

Elderly entitlement programs the world over bind generation to generation through the mediation of government. The typical elderly entitlement program is financed by generation transfers, in which government taxes the young and transfers the proceeds to retirees. In the United States, retirees rely on the young for their Social Security pensions. In exchange for their taxed away earnings, the young are given the implicit promise that they, too, will receive a pension in their retirement. One of the intrinsic defects of elderly entitlement programs financed with generation transfers is that their financial health is very sensitive to demographic changes. The retirement of the baby boom generation will usher in a decreased worker-to-retiree ratio. Combined with increased life expectancies and reduced birth rates, the falling worker-to-retiree ratio is expected to continue for the foreseeable future. These demographic shifts will place elderly entitlement programs in deep financial crisis. Taking Social Security and Medicare in the United States for example, scheduled tax revenues will fall dramatically short of the resources required to pay promised benefits. Fundamentally, the problem of financial insolvency of these programs lies in the fact that they are financed by intergenerational transfers rather than by resources based on saving and investment.

Given the fact that generation transfer systems in effect throughout the developed world will be in serious deficit early in this century, the potential exists for a significant increase in the share of total income that passes through government. In the United States, for example, if benefits for Social Security and Medicare are paid as currently scheduled, the government's share of gross domestic product will increase from its current level of 18 percent to 37 percent within a generation. By way of comparison, the federal government was 10 percent of gross domestic product prior to World War II, rose to 45 percent in 1944, but fell back to 14 percent following the cessation of hostilities. The increase in the share of total production that must pass through government as a result of generation transfers has the potential of being the next great usurper of private property.

The looming financial crisis of elderly entitlement programs can be resolved without creating more government interference in the economy by transforming these programs into retirement systems of privately owned savings accounts. The benefit from a change to a prepaid retirement system occurs in the long run, whereas the burden of the change would fall largely on current workers and near-future generations. Why such a change would be desirable and how such a change could be realized are the topics of this paper.

The main benefit of a transition from publicly provided old age pensions to privately owned retirement accounts comes from the increase in the nation's capital stock as a direct result of the transition. Thus, a necessary, but not sufficient, condition for supporting the abandonment of our current pay-as-you-go system of financing elderly entitlements in favor of a system of private accounts is that in the long run both the retired and working generations will enjoy greater consumption. This condition will be satisfied after the transition because all generations that come after the completed transition will be free from any debt implicit in a transfer-financed public pension system.

Assessing the cost of transition that must be borne before we reach the point where all post-transition generations are completely relieved of the implicit debts is not as straightforward as it may seem. Any reform must be compared with a benchmark that is itself sustainable. Social Security and Medicare, with the existing benefit and tax schedules, cannot serve as such a benchmark because neither program, without significant benefit cuts or tax hikes, is financially solvent. The real costs of transition are those that transitional generations must bear that are above and beyond the sacrifice they would have to make to maintain a solvent transfer-based entitlement program. Nevertheless, since both the real costs of transition and the costs that must be incurred to bring about a sustainable generation transfer system have to be paid by the transitional generations, the sum of both is often referred to as the transition cost.

As of January 1, 2003, the existing members of the Social Security system are owed a debt of $11.9 trillion. This debt, equal to the present value of benefits in excess of tax payments, must either be paid or reneged upon if future generations are to be put in a new retirement system based on privately owned accounts. However, even absent any transition to a prepaid retirement system, the $11.9 trillion Social Security debt exists and must be paid, again through a combination of increased taxation and/or benefit reductions. In a sense, a transition to benefit prepayment does not generate any additional costs but only brings forward the pain of paying off the existing debt.

In all reform proposals that envision a transition to a retirement system based on privately owned savings accounts, the fundamental issue is how the transition costs should be distributed among transitional generations. In this paper, we study three transitions from the current pay-as-you-go system of financing elderly entitlements to a system of private accounts. In our analysis, we focus on aggregate quantities and intergenerational equity, therefore implicitly treating individuals of the same generation as identical. A concern expressed by opponents of Social Security privatization has been that general private individual account retirement systems, such as the ones presented in this paper, tend to be less redistributive than current public systems.[1] While intragenerational equity is not a consideration in this paper, the issue of intragenerational redistribution can be handled within a system of individual accounts where the aged poor are treated in a manner similar to the non-aged poor.[2]

The remainder of this paper is organized as follows. First, we introduce some simple financing identities of generation transfer programs and their implications, followed by a general discussion of the benefits and costs of converting a generation transfer program into a prepaid system with private savings accounts. Then, using prepaying current U.S. Social Security as an example, we analyze various aspects of a transition.

SOME SIMPLE ACCOUNTING OF GENERATION TRANSFER SYSTEMS

To understand some of the constraints of intergenerational transfer programs, we present below some simple financing identities of generation transfer systems. For this purpose, we divide the current and the future population into two groups: the "closed group," consisting of the current adult population (those fifteen years and older), and the "new group," consisting of the current pre-adult population and all yet-to-be-born generations. At the same time, we shall refer to the union of these two groups (all current and future generations) as the "open group."

Denote the present time as t_0. For any intergenerational transfer program, since the closed group and the new group do not intersect, open group income (OGI) at any time $t \geq t_0$ can be expressed as the sum of closed group income (CGI) and new group income (NGI). In the same manner, open group expenditure (OGE) at that same point in time can be expressed as the sum of the closed group expenditure (*CGE*) and new group expenditure (*NGE*). Thus,

(1) $$OGI_t = CGI_t + NGI_t,$$
$$OGE_t = CGE_t + NGE_t.$$

Define the open group unfunded obligation (*OGUO*) at time t_0 as the present value of the difference in open group expenditure and open group income from the same point in time into the indefinite future. We have then that

(2) $$OGUO_{t_0} = \sum_{t=t_0}^{\infty} \frac{OGE_t - OGI_t}{(1+r)^t},$$

which from (1) can be written as

(3) $$OGUO_{t_0} = \sum_{t=t_0}^{\infty} \frac{CGE_t + NGE_t}{(1+r)^t} - \sum_{t=t_0}^{\infty} \frac{CGI_t + NGI_t}{(1+r)^t} = \sum_{t=t_0}^{\infty} \frac{CGE_t - CGI_t}{(1+r)^t} + \sum_{t=t_0}^{\infty} \frac{NGE_t - NGI_t}{(1+r)^t}.$$

Defining the first and second summations on the final right-hand side of (3) as the time t_0 closed group unfunded obligation ($CGUO_{t_0}$) and new group unfunded obligation ($NGUO_{t_0}$), respectively, we have

(4) $$OGUO_{t_0} = CGUO_{t_0} + NGUO_{t_0}.$$

The above open group unfunded obligation, calculated as the present value of present and future scheduled expenditures less scheduled income, is often referred to as the infinite horizon financing shortfall of a transfer-financed entitlement program. As a component of the $OGUO_{t_0}$ and calculated as the present value of the difference between closed group expenditures and income, the closed group unfunded obligation is referred to by the Social Security and Medicare trustees as the 100-year closed group debt because of its similarity to government debt that is held by the public.[3]

Define a sustainable generation transfer system as one with an open group unfunded obligation of zero. When we begin the discounting process, the closed group contains all the current taxpayers and transfer recipients. As the system ages, the proportion of the closed group that provides income to the system falls as taxpayers become transfer recipients. Thus, even in a sustainable system, the closed group unfunded obligation is always positive. Therefore, in a sustainable system we know from (4) that the new group unfunded obligation must be negative and equal in absolute value to the closed group unfunded obligation. Not a surprising result, since for all $t > t_0$ and less than the age at which benefits are paid, the new group contains no recipients, only taxpayers. We shall use this simple arithmetic of generation transfer systems later.

In general, because of a worldwide boom in population that occurred in the period often dated from 1946 to perhaps 1964, social security systems around the world have scheduled tax rates that are below the tax rate that will be required to pay future scheduled benefits. The existence of this baby boom may well have led Paul Samuelson to say, "Social Security is squarely based on what has been called the eighth wonder of the world—compound interest. A growing nation is the greatest Ponzi game ever contrived. And that is a fact, not a paradox." (*Newsweek*, February 13, 1967)

Since the baby boom did not continue, at scheduled tax rates and benefits, newcomers to the system will contribute little or nothing to pay off the debt owed to the closed group. As an example of this fact, we show in Table 1 the three unfunded obligation measures defined above for the present U.S. Social Security system based on the 2003 Trustees Report.[4] As the table indicates, the U.S. system, similar to all other retirement systems in the developed world, has a long-run problem indicated by the fact that the new entrants to the system will provide no resources to pay off the closed group debt. Whether or not the system is reformed, the debt owed to the closed group must be either paid or canceled. If current members are allowed to receive promised benefits while paying only scheduled taxes, new members' taxes must be raised because the retirement of current members will consume real resources.

For a system that is financially sustainable, there would be no financing shortfall, and therefore, the 100-year closed group debt would be paid off by new entrants to the system so that

$$(5) \qquad CGUO_{t_0} = -NGUO_{t_0}.$$

This constraint on any solvent entitlement program highlights the zero-sum nature of closed group debt financing: If we want to reduce the financial burden on future participants (the new group), the debt owed to current participants (the closed group) must be partially revoked, either by reducing the benefits or increasing the taxation of closed group members. The fact that the new group unfunded obligation is approximately zero indicates something else that may not be obvious: At the current tax rate, if the surpluses in the early years of the new group were invested at the assumed discount rate, the resulting fund would be sufficient to pay scheduled benefits. Thus, while the tax rate is not sufficient to fund a generation transfer retirement system, it is sufficient to fund a prepaid retirement system.

Table 1
U.S. Social Security System Financing Shortfall and Its Decomposition
(Present values as of Jan. 1, 2003, in trillions of dollars)

Open group unfunded obligation (financing shortfall)	$11.9
Closed group unfunded obligation (100-year closed group debt)	$11.9
New group unfunded obligation	$0

THE BENEFITS OF PREPAYING WITH PRIVATE ACCOUNTS

Several reasons have been put forward in favor of prepaying Social Security with private savings accounts. The first and foremost among these is that, absent a contract with unborn generations, members of a cohort must provide for retirement by storing output, essentially acquiring capital, during their productive years. Thus, the movement to advance funding will increase the capital stock relative to its current level and allow future generations to earn higher income. The huge debt implicit in the current retirement system, in the form of accrued benefits, has replaced this need to acquire capital so that future generations inherit a smaller capital stock and have lower income. Essentially, a transition to prepaid retirement benefits with private accounts will bring forward the debt-servicing schedule and hence, increase the nation's capital stock. The debt-retirement benefits from a transition to prepaid benefits will be discussed in detail in a later section where three transition paths for Social Security are simulated.

A second reason for prepaying with private savings accounts is to resolve the current programs' financing crisis without increasing government's share in the economy in the form of publicly operated intergenerational transfers that bind one generation to another. But, one might ask, why is it undesirable to bind generations together through government? In the context of reforming Social Security, it is appropriate to rekindle some of Thomas Jefferson's thoughts as they relate to binding one generation to another. Jefferson was a champion of freedom, and his intellectual curiosity led him to comment on a broad array of topics, including Social Security. Well, maybe not Social Security in particular, but in an intriguing letter to James Madison, Jefferson develops the proposition "that the earth belongs in usufruct to the living: that the dead have neither powers nor rights over it."

Though this proposition may appear on the surface to have no direct application to a generation transfer program such as Social Security, Jefferson's development of the concept reveals a keen insight into the problems and the philosophical implications of binding one generation to another via long-term debt. In terms of debts, Jefferson states in the same letter, "Then no generation can contract debts greater than may be paid during the course of its own existence." Jefferson's logic was that if a generation could leave a debt to the next generation upon its death, "then the earth would belong to the dead and not the living generation."[5]

Social Security binds one generation to another by always leaving a debt to the incoming generation. The debt is the implicit promise to pay benefits to retirees. Each retiree holds an implicit bond equal to the expected present value of his or her benefits. It is the substitution of these implicit bonds for capital that represents the true cost to society of generation transfer-based retirement systems. By endowing the initial generation of beneficiaries with pensions, the continuation of the system results in each new generation inheriting a debt. The debt is never fully retired by the working generation, for as its members pay for the benefits of retirees, they accrue benefits of their own that become debts of the next generation.

A third and often argued reason for prepaying is that it can reduce the welfare loss due to payroll taxes by eventually lowering or eliminating the tax. The negative incentive effects of the payroll tax have been used in the reform debate to argue that it may be possible to have a Pareto-improving transition of the current public Social Security system to a prepaid system with private savings accounts in the sense that all the living and future generations are made better-off. Absent the possibility of a Pareto transition, some arbitrary relative value of the welfare of current versus future generations is implied by a move to private markets.[6] Several studies have claimed that such a Pareto-improving transition is not possible,[7] whereas others have found such Pareto-improving transition paths through simulation and attributed the sources of these "win–win" transitions to some sort of preexisting distortion in the economy.[8]

In Liu, Rettenmaier, and Saving (2000), we used an analytical framework with both labor and capital market distortions to investigate the possibility of a Pareto-improving transition. We found that when the links between Social Security payroll tax contributions and Social Security benefits are sufficiently weak, privatization will yield a Pareto-improving efficiency gain by simply replacing the implicit debt with explicit debt without increasing the nation's capital stock. In essence, the Social Security debt has to be serviced with or without private savings accounts, but the issue is that the current system links this debt-servicing tax to payroll, while under a reformed system, a general tax would play this role. As we know, the payroll tax is bad in that it punishes productive behavior. With the payroll tax replaced by the less distortionary general tax, there can be an efficiency gain in which every generation can be made better-off.[9] However, since one could always replace an inefficient tax with a more efficient one, such a change should not be a benefit of the transition.

Finally, one may think that the benefits of prepaying can be achieved by government investing a trust fund in the capital market rather than through the establishment of private accounts. This is doubtful since for one thing, the government has never been able to do so. Today, in fact, the relatively modest Social Security Trust Fund consists entirely of Treasury IOUs. Even if it were possible for the government to commit to investing in real assets, giving the federal government the green light to invest in our nation's equities would raise a number of issues concerning the separation of the government and the private sector, with the danger of politicizing firm decisions.

THE COSTS OF THE TRANSITION

It does not necessarily follow that every generation would be better-off with a transition from an existing public pay-as-you-go elderly entitlement system to a system of private individual accounts, even though a significant increase in lifetime consumption could occur in the long run after the reform. The key to understanding the necessary sacrifice that must be made during a transition is the "transition cost" that must be incurred to deal with the debt implicit in the promised benefits of the existing pay-as-you-go system.[10]

Any transition from a generation transfer system of retirement to one in which individuals in each birth cohort provide for their own retirement must deal with the debt of the old regime. As we show above, this debt—the closed group unfunded obligation—can be measured by calculating the present value of net benefits to existing generations. This calculation considers everyone currently in the system and allows them to remain in the current system. The debt owed to this group is referred to as the 100-year closed group debt and has been estimated to be $11.9 trillion in 2003. We have also shown that, in a generation transfer system that is financially solvent, the net contribution of new entrants will exactly equal the 100-year closed group liability. In fact, the projected net contribution of new entrants to the U.S. Social Security system is essentially zero.

This unfunded retirement system debt is only partially the result of pay-as-you-go financing. The larger-than-normal baby boom working generation also plays a role because its tax rate, while sufficient to fund the retirement of a relatively small retired generation, will be woefully insufficient to fund the retirement of the large baby boom generation. For example, the current U.S. Social Security system provides retirement benefits equal, on average, to an income replacement rate of 42 percent. The trustees estimate that as early as 2030, there will be only two workers for each retiree, which implies, in equilibrium, a tax rate of 21 percent. In contrast, the current tax rate is 10.7 percent, just over half the tax rate that will ultimately be required.

To put this problem in perspective, let us combine the two major elderly entitlement programs in the United States. Table 2 shows the resulting financing shortfall for Social Security and Medicare and its decomposition between the closed and new groups. The present value of accrued benefits owed existing members of the system is $24.4 trillion. The present value of the scheduled net cost of newcomers is $25.3 trillion, making total unfunded obligations almost $50 trillion, more than fifteen times the acknowledged federal debt.[11]

One way to deal with an accrued elderly entitlement debt is to bite the bullet and raise taxes immediately by an amount sufficient to amortize the entirety of future promised benefits. This approach requires a tax rate greater than the actuarial deficit reported by the trustees because their actuarial deficit is only adequate to take the system 75 years into the future. Since the system is in substantial deficit at the end of 75 years, a much greater tax increase would be required to ensure solvency in the long run. This year, for the first time, the trustees reported the perpetuity actuarial deficit for Social Security. At the

Table 2
U.S. Social Security and Medicare System Financing Shortfall and Its Decomposition
(Present values as of Jan. 1, 2002, in trillions of dollars)

Open group unfunded obligation (financing shortfall)	$49.7
Closed group unfunded obligation (100-year closed group debt)	$24.4
New group unfunded obligation	$25.3

assumed 3 percent real discount rate, the trustees report that a once-and-for-all increase in the tax rate of 3.8 percentage points would make the U.S. Social Security system solvent forever. However, such an approach is doomed to failure if the increased taxation is not accompanied by investment in the real economy of the early surpluses generated by the new taxation, coupled with a real property right assignment of this new capital.

Given that private savings accounts have to be established for the new entrants to the workforce, the choice among alternative transition paths is simply an issue of intergenerational equity: how the burden of the $11.9 trillion Social Security debt—the burden of the transition costs—should be distributed among present and future generations. In one extreme case, the current U.S. Social Security system could be replaced by letting new entrants have private accounts and allowing all those currently in the system to remain and receive full promised benefits and pay existing tax rates. For this transition, new entrants would have to provide the entire $11.9 trillion to existing generations in addition to providing funds for their own retirement. Any transition to a system of private accounts, while maintaining scheduled benefits and taxes, leaves currently living generations with the status quo but makes future generations worse off. If we are to succeed in finding a transition that has any potential for being intergenerationally equitable, we must include some or all of the current generations in the transition.

Consider another extreme case where the current U.S. Social Security system is replaced by letting new entrants have private accounts and giving all those currently in the system recognition bonds worth a total of $11.9 trillion. Assume a new consumption tax is raised to service the debt. In doing so, all the living generations, including both current workers and retirees, share with the future generations in paying for the burden of financing the $11.9 trillion Social Security debt. In essence, the debts owed to both current workers and current retirees are partially reneged. However, a transfer-based pension arrangement tends to take on a life of its own and is extremely hard to make smaller or to eliminate. It is hard to take benefits away from current retirees, given they have reduced earnings capacity and have adjusted their savings behavior in light of the expected transfers. It is also hard to reduce the scheduled benefits of near-term retirees because they have also already planned their lifetime savings behavior assuming that Social Security would be there for them. Therefore, it may be argued that dealing the current and the soon-to-be retirees an unexpected financial blow may be politically infeasible.

Between these two extremes, a middle-of-the-road alternative is to let the current retirees and those workers fifty-five years and older remain in the current system by having them receive scheduled benefits and pay scheduled taxes as well. In contrast, private savings accounts would be established for workers fifty-four years and younger and new entrants as they come. At the same time, both the young workers and the future entrants will pay for the phaseout of the current system until all those who are currently fifty-five years and older have exited the current system.

The approach taken by most reformers is a variant of the alternative transition paths discussed above. These reform plans require future (and in some cases current) workers to establish private savings accounts for their retirement expenses, give up the right to some or all of their generation transfer benefits, and pay taxes sufficient to support current and soon-to-be retirees. Assuming that desired retirement income is greater than or equal to the future value of the new mandatory savings accounts, this new saving will result in additions to the capital stock and increased national income.

The increased national income will eventually allow for increased consumption. Feldstein and Samwick (1997), and subsequently Feldstein, Ranguelova, and Samwick (1999), have suggested that all current workers establish mandatory personal retirement accounts (PRAs) and continue paying payroll taxes at the current rate. Initially, the contribution rate to the private accounts, as a percentage of wage earnings, is low—in the range of 2 percent. As funds accumulate in the private accounts, two things happen. As the system matures, the annuities that can be purchased at retirement offset an increasing proportion of scheduled Social Security benefits, thus reducing the financing requirements of the current system.

Assuming that they are required to pay the full cost of paying retirees benefits, current workers bear a greater burden than they would under the pay-as-you-go system, but future workers would be much better off under a prepaid system than under the current system. So, such a transition is not necessarily Pareto-improving. The gradual reduction in the payroll tax will reduce the deadweight loss due to the reduced labor supply under the current tax rate. Feldstein and Samwick (1997) estimate the efficiency gains from such tax rate reduction to be about 2 percent of the tax base.

Kotlikoff and Sachs (1998) have offered another transition path. Focusing exclusively on the retirement portion of Social Security, they suggest eliminating the payroll tax and replacing it with mandatory contributions to private accounts. The transition cost associated with the accrued benefits would be financed by a new federal business cash flow tax. Since both retirees and workers engage in consumption expenditures, the tax burden for the transition is shared by both workers and retirees. In addition, the business cash flow tax is less distortionary than the payroll tax, and therefore, the switch in the tax by itself produces an efficiency gain. Over time, the tax rate associated with the new cash flow tax would decline as the liabilities of the phased-out system are eliminated.

The President's Commission to Strengthen Social Security (2001) suggested that private accounts offset some of the accumulated debt. More important, however, the commission also suggested that rather than replace a constant share of wage-indexed earnings, the defined benefit part of Social Security provide a fixed level of purchasing power. This change alone reduces the outstanding debt to existing generations, the 100-year closed group liability, significantly reducing the tax required and reducing the cost to both new and existing generations. Such a change is one way of recognizing the fact that the existing system is not sustainable and, therefore, is not the appropriate target when deciding whether a transition is Pareto.

For example, by 2021, just four years after the trustees forecast that Social Security revenues will fall short of benefit payments, the Treasury will have to transfer to Social Security the equivalent of 5 percent of all projected federal income tax revenues. Historically, the largest such transfer has been 4.5 percent of federal income tax revenues (in 1978 and 1983). In both these years scheduled benefits were cut, and in 1983, taxes were also raised. Figure 1 shows the transfers as a percentage of total projected federal income tax revenues that will be required to pay scheduled benefits for Social Security and both parts of Medicare, based on the 2003 *Annual Report of the Board of Trustees of the Federal Old-Age and Survivors Insurance and Disability Insurance Trust Funds*. Together with Medicare, the Treasury will, in 2020, be transferring to the elderly the equivalent of 17.5 percent of total income tax revenues and by 2030, 36.5 percent of all federal income tax revenues. In contrast, this year, these programs contributed to the Treasury revenues equal to 0.5 percent of total income tax revenues. Thus, to require that all current participants receive full scheduled benefits, in assessing whether or not the transition to private accounts can be made on a Pareto-improving basis, is probably an unfair requirement.

THREE TRANSITIONS TO FULLY FUNDED SOCIAL SECURITY

The preceding discussion identified several possible transitions from the current pay-as-you-go financing of elderly entitlements to a system of private accounts that prepay some or all Social Security benefits. In the discussion that follows, we present three transitions based on the current U.S. Social Security system.

Figure 1
General Revenue Transfers to Social Security and Medicare
(as a percentage of income taxes)

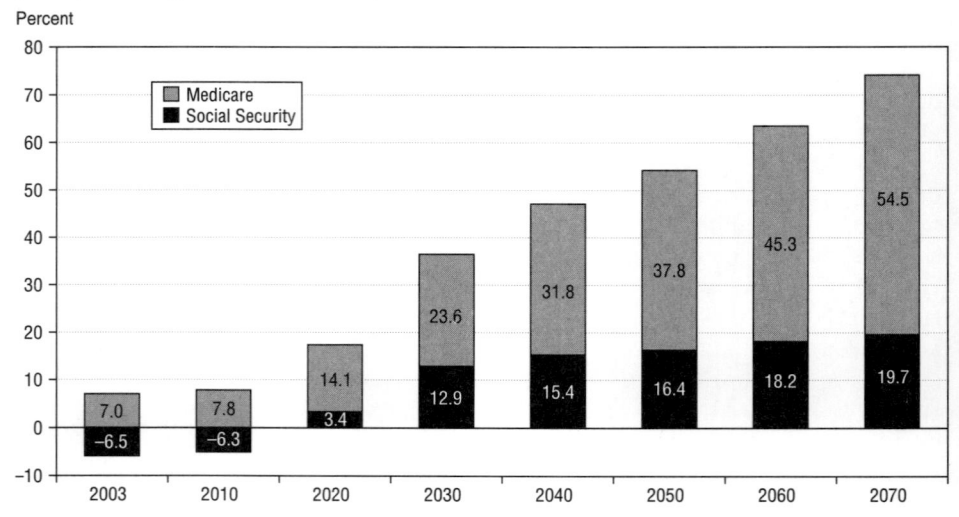

Percent

Year	Social Security	Medicare
2003	−6.5	7.0
2010	−6.3	7.8
2020	3.4	14.1
2030	12.9	23.6
2040	15.4	31.8
2050	16.4	37.8
2060	18.2	45.3
2070	19.7	54.5

The first transition, **Reform I**, keeps intact the benefit structure for all those currently in the system (those fifteen years and older) but lets the tax rate be constantly adjusted so that revenues are always equal to expenditures, that is, the tax rate is set at what the trustees refer to as the cost rate. By letting those currently in the system pay the cost rate, they participate in paying some of the 100-year closed group liability.

At an assumed rate of return of 5.4 percent (the return on a 60 percent equity, 40 percent bond portfolio over the past sixty years), the required contribution rate to yield a 42 percent replacement rate is 4.22 percent. We assume that all individuals under age sixty-seven pay the tax rate required to pay scheduled benefits—the trustees' cost rate. We further assume that all new entrants to the system get no benefits from the old system, are required to place 4.22 percent of their income into a private account and pay the cost rate that is necessary to pay benefits to the closed group. As the population eligible for generation transfer benefits falls, the tax rate will begin to decrease and reach zero in 100 years, when the last of those currently in the system are expected to be deceased.

Figure 2 contrasts the cost rates with this reform to the cost rates necessary to maintain the status quo pay-as-you-go financing. The first two series to consider are the status quo cost rate and the closed group cost rate. The status

Figure 2
Income and Cost Rates for the 100-Year Closed Group and for Future Generations with a Reform That Pays Off 100-Year Closed Group Debt
Percentage of taxable payroll

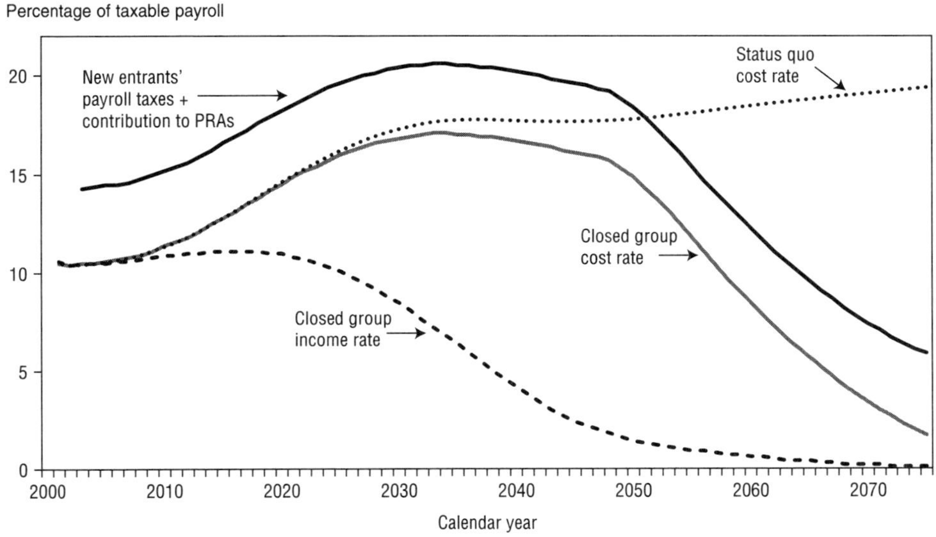

SOURCE: 2001 Social Security Trustees Reports and SSA 2001 Cohort Operations file.

quo cost rate reveals the prominent increase in spending as a percentage of taxable payroll between 2010 and 2030 that is associated with the retirement of the baby boom generation. The status quo cost rate continues to rise after 2030 but at a slower rate. The closed group cost rate shows the share of the status quo cost rate that is attributable to the members of the closed group: individuals who are fifteen and above today. As this series indicates, the cost of making benefit payments to this group rises to 17 percent of payroll by 2034 and then begins to fall. By 2075, the cost is about 2 percent of payroll, and 100 years out, the cost has dropped to zero. The difference between the status quo and closed group cost rate is the costs associated with new entrants to the system.

The top cost rate shows the time path of the new entrants' combined payroll taxes and their contributions to private accounts. Their taxes and contributions start at 14.3 percent of taxable payroll and rise to 20.6 percent by 2034 and then decline until they ultimately fall to the contribution rate required to fund their personal retirement accounts. The figure shows that until 2052, the combined costs for newcomers are in excess of what they would have paid under the status quo. The final line in the graph shows the total taxes paid by the closed group represented as a percentage of taxable payroll. Currently, the closed group accounts for 100 percent of all taxpayers, but as time goes on, members of the closed group retire and stop paying taxes, as reflected by their declining income rate.

Table 3 shows three estimates of the present value of the 100-year closed group expenditures and income for the existing Social Security system and the **Reform I** program in which members of both the closed group and the new group pay the cost rate. The closed group debt, as measured by the trustees and shown in the column labeled "conventional" in the table, is $11.9 trillion. However, this debt treats the surpluses generated between 2003 and 2017 as if they are invested in real assets. In reality, these surpluses are simply spent by Congress and no investment occurs. Thus, the correct measure of the debt, with the surpluses eliminated, is $13 trillion, as shown in column 2 of the table. Finally, the reformed system, in which the closed group pays the cost rate, has a closed group debt of $11.8 trillion. Thus, by requiring the closed group to pay the cost rate that is necessary to pay their benefits, the group pays off some of its debt with **Reform I**, but it still leaves a little over 90 percent of the debt to new entrants.

Under **Reform I** the first newcomers are worse off, since their taxes plus contributions are higher than would have been required to keep the generation transfer system intact throughout their work life. Further, it is not until 2060 that total taxes by the new group are less than the current legislated tax rate. However, it should be emphasized that this legislated tax rate is insufficient to keep the generation transfer system solvent. This first example of a total prepayment reform illustrates two significant aspects of any reform. First, it illustrates the share of the closed group liability that, assuming no changes in scheduled benefits, can be paid by having the closed group's taxes rise with the cost of pay-

Table 3
Three Estimates of the 100-Year Closed Group Unfunded Obligation in 2003

Category	Conventional	Surpluses Not Included	Reform I
Present value of revenues	13.9	12.8	14.0
Present value of expenditures	25.8	25.8	25.8
Remaining closed group debt	−11.9	−13.0	−11.8

SOURCES: Social Security Administration 2001 Cohort File and 2003 Trustees Report. The column titled "Conventional" is the standard way of calculating the obligation. The next column sets the income rate to the cost rate between 2003 and 2017. The third column does not include the surpluses either, but it does impose the closed group cost rate on members of the closed group beginning in 2018. None of the estimates include the Trust Fund offset as a revenue source.

ing their benefits. Second, it illustrates the burden placed on new entrants if they are required to both pay off the remaining closed group debt and, at the same time, prepay their own retirement.

A significant shortcoming of **Reform I** is that it treats members of the closed group and newcomers differently. It is unlikely that explicit differential "tax" rates in any given year, such as those in the previous example, would be acceptable to taxpayers. Most reform proposals envision either identical mandatory contribution rates to private accounts or equal percentage contributions paid from one's payroll taxes. Either way, all taxpayers are treated the same in a given year. Reforms that include private accounts can also take the form of either partial or total prepayment.

We turn now to analyzing a partial prepayment and another full prepayment reform. An example of a partial prepayment reform is the second reform put forward by the President's Commission to Strengthen Social Security, which we will refer to as **Reform II**. That reform would allow workers to contribute roughly 31.7 percent of their payroll taxes, 4 percentage points of the total 12.6 percent payroll tax, up to $1,000 per year, to a private account. With the restriction of $1,000, the total contributions to private accounts ultimately reach 2.39 percent of taxable payroll. The reform also replaces the wage-indexed benefits formula with a price-indexed formula beginning in 2009. Price indexing effectively sets the defined benefit after 2008 to the real purchasing power of the 2008 benefit.[12] In exchange for the opportunity to divert one's payroll taxes to a private account, the price-indexed benefit is offset by the annuity resulting from one's private account, assuming the private account earns 2 percent. Future benefits are first reduced by the new benefit formula and are then further reduced by the benefit offset, assuming the 2 percent rate of return. Any accumulations earned in excess of the 2 percent are added to the reformed benefits to arrive at a retiree's total benefit. Using the commission's assumptions, the ultimate benefits from this reform are roughly comparable to those currently scheduled.

In the calculations reported below, we assume 100 percent participation and that any funding shortfalls are made up using payroll taxes. While the shortfalls could be financed in any way—for example, income taxation—we use payroll taxation to make the results of this reform comparable to **Reform I** discussed immediately above. The use of payroll taxation makes the closed group pay less of their liability than general income taxation because, with income taxation, even the retired population participates in paying off the debt.

In our second example of a full prepayment reform, **Reform III**, we require all future participants and all current participants fifty-five years of age and younger to contribute 4.22 percent of their payroll to a private account. In addition, both future and current participants pay payroll taxes equal to the cost rate. With this contribution rate, the annuity that a new entrant could purchase at retirement would be roughly equivalent to scheduled benefits.[13] At the same time, scheduled benefits are reduced by the expected value of the annuity that this contribution rate would purchase.[14] In this way, individuals in the closed group fifty-five years of age and younger prepay part of their retirement pensions. It should be emphasized that the average annuity that can be purchased using the personal retirement account accumulations identifies the benefit reduction schedule for **Reform III**. This benefit reduction schedule is pre-announced and is part of the reform and is thus similar to the pre-announced change to the price-indexed benefit formula in **Reform II**. Each successive cohort knows at the beginning of the reform the expected size of their tax-financed defined benefit.

Figure 3 illustrates the cost rates resulting from the President's Commission, **Reform II**, and the second full prepayment reform, **Reform III**. The two reforms' cost rates shown in Figure 3 illustrate how each reform reduces cost relative to the cost of paying scheduled benefits. As expected, the larger contribution rate leads to a more dramatic and rapid reduction in the cost of the pro-

Figure 3
Cost Rates for Status Quo and for Two Reforms When All Costs Are Paid Using Payroll Taxes

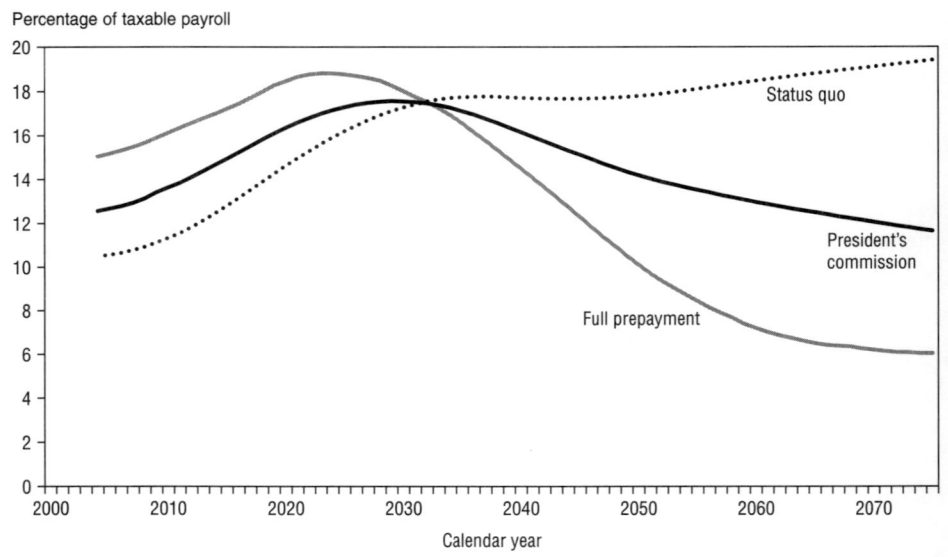

Percentage of taxable payroll

SOURCES: 2001 Social Security Trustees Reports and SSA 2001 Cohort Operations file, the President's Commission income and cost rates are based on the second plan from the 2001 report of the President's Commission to Strengthen Social Security, and the full prepayment income and cost rates are based on authors' estimates.

Table 4
Effects of Three Reforms on the Closed Group Obligation

Category	Surpluses Not Included	Reform I	Reform II President's Commission	Reform III
Present value of revenues	12.8	14.0	13.4	12.6
Present value of expenditures	25.8	25.8	22.0	18.5
Remaining closed group debt	−13.0	−11.8	−8.6	−5.9

SOURCES: Social Security Administration 2001 Cohort File and 2003 Trustees Report. None of the estimates include the Trust Fund offset as a revenue source.

gram that would have to be paid through taxes. In fact, the share of payroll going to program costs and private accounts is 6.05 percent in 2075 for **Reform III**, the second full prepayment reform, and 11.66 percent for the President's Commission reform, **Reform II**. Both reforms have total costs that exceed the status quo up to 2032 and then become increasingly better than the status quo.

Table 4 presents the effects of **Reform I**, the President's Commission reform, **Reform II**, and **Reform III** on the 100-year closed group obligation. The first two columns are from Table 3 and provide a reference point for the other two reforms. The commission proposal reduces the closed group net obligation by $4.4 trillion, about 34 percent, primarily by reducing benefits. Our assumption that all shortfalls are covered through payroll taxes results in a small increase in closed group revenues for the commission proposal of $0.6 trillion. Not surprisingly, **Reform III** has the greatest effect on the share of the closed-group debt paid by the closed group as it reduces the closed group debt by $7.1 trillion, or 54 percent.

In all three reforms, how the deficits are financed determines the degree to which they produce changes in the capital stock. For a reform to increase the capital stock, the implicit debt must be reduced. This means that reforms must be debt reducing to produce beneficial capital stock effects. Financing any reform with debt means that total debt remains unchanged and no capital stock effect occurs as individuals continue to use debt rather than capital to transfer resources across time.

The choice of the tax instrument used to pay the initial burdens of the reforms would also have economic ramifications. A broadly based tax, such as a consumption tax, does two things. It has a smaller deadweight loss than a payroll tax that raises the same revenues. It also spreads the burden of the tax to retirees. This second point is important as we look at the timing of the additional burden of each reform. Given that the baby boomers have paid lower lifetime Social Security taxes than will be required of the next generation, in a generational equity sense it could be argued that the baby boomers should share in the cost of prepayment. This sharing of the cost could be accomplished by a transitional consumption tax during their retirement.

CONCLUSION

This paper provides an analysis of the benefits and costs of a transition from intergenerational transfer financing of elderly entitlements to intragenerational financing of these same entitlements. The benefit side of the equation is a result of the increase in the capital stock and income that, after the transition, translates into increased consumption for all future generations. The cost side reflects the necessary reduction in consumption for the younger generations during the transition period. We simulate three transitions to a prepaid system of elderly entitlements based on the current U.S. Social Security program.

As the simulations indicate, the more complete the reform, the higher are the initial costs and the higher are the long-run benefits in terms of the degree to which the payroll tax is reduced and the degree to which the capital stock increases. Ultimately, reforming how Social Security is financed is a political decision. Since the next generation has no voice in the decision, reforms with long-term benefits will be discounted in the voting process.

The estimates presented assume particular transitions to a private system of providing for elderly retirement benefits. There are other approaches, all of which can accomplish the goal of moving us from generation transfer-based Social Security to prepaid Social Security. Fundamentally, however, the financing issues addressed here must be faced whether or not any change is made in the basis of Social Security financing. Admittedly, the financing issue can be solved by providing those currently in the system with reduced benefits and increased taxes. Such a transition will leave the new members with a smaller debt and allow them to have more consumption than the transition discussed here. No matter how we make the transition, the elderly are going to consume real resources, and as the elderly population grows, the younger generations are going to have to give up consumption in favor of the elderly. The only real question is how these younger generations will be induced to give up the resources necessary to provide the elderly with their retirement benefits.

NOTES

[1] The current system is less progressive than it might seem from its highly redistributive benefit schedule due to a positive correlation between lifetime income and longevity. According to Gar-

rett (1995), differences in mortality considerably narrow, and in some cases eliminate, the progressive spread in returns across income classes. Liu and Rettenmaier (2003) also reached a similar conclusion by studying both the rate of return and the present value of the Social Security investment for different racial and education groups.

[2] For a detailed analysis of how individual accounts and intragenerational redistribution can be mutually compatible with progressive matching of individual accounts, see Kotlikoff, Smetters, and Walliser (1998).

[3] Current mortality tables imply that almost all of the existing population of 15-year-olds will be deceased by age 115. Thus, in 100 years, the closed group essentially contains no members.

[4] See 2003 *Annual Report of the Board of Trustees of the Federal Old-Age and Survivors Insurance and Disability Insurance Trust Funds*.

[5] In this same letter, Jefferson went on to state, "On similar ground it may be proved that no society can make a perpetual constitution, or even a perpetual law. The earth belongs always to the living generation." All excepts are from *The Writings of Thomas Jefferson*, H. A. Washington, editor (1853–54, 392–97). Madison responded to this letter by noting that some investments on the part of the government, such as conducting a war, have far lasting benefits that may justify future generations' participation in paying for the investment.

[6] Many privatization proposals in the United States have adopted immediate tax increases as a way of financing the transition cost. As admitted in these proposals, however, it is often the case that the long-run benefits of a transition financed by sharp immediate tax increases come only at a cost to the initial working generations. Using a criterion of discounted present value, some studies have claimed an overall efficiency gain from this type of transition (see Feldstein 1995). Such a comparison between gains to one generation and costs to another generation, however, must resort to an across-generation welfare function.

[7] For example, see Geanakoplos, Mitchell, and Zeldes (1998), Mariger (1997), and Murphy and Welch (1998).

[8] For example, Feldstein and Samwick (1997) argue that the possible Pareto improvement from Social Security privatization comes from reduced capital market distortion. On the other hand, Kotlikoff (1998) identifies the main source of the efficiency gain from privatization to be the labor market distortion caused by payroll taxes.

[9] Therefore, any Pareto improvement that is the result of a transition from pay-as-you-go Social Security to prepaid retirement can be accomplished by a tax reform without any change in the generation transfer-based financing.

[10] In fact, prepayment does not generate any additional cost—at least not the kind of cost captured by this term—but serves only to bring the implicit government debts, in terms of accrued benefits in the pay-as-you-go system, to the surface.

[11] For a detailed discussion of the unfunded obligations in both Social Security and Medicare, see Liu, Rettenmaier, and Saving (2003). Note that the estimates in Table 2 are for 2002 rather than 2003.

[12] Given positive income growth, fixing the real defined benefit makes this reform a total prepayment reform in the limit as the ratio of prepaid benefit to defined benefit goes to zero.

[13] This contribution rate combined with a 5.4 percent rate of return would replace approximately 42 percent of an average worker's earnings. It has been argued that prepayment proposals should use lower rates of return either associated with a risk-free asset such as inflation-indexed bonds or on a financially engineered instrument that guarantees the pension return. This would be appropriate if scheduled Social Security payments were themselves guaranteed, but history shows that the Social Security "investment" changes over time in terms of the tax rate, taxable maximum, benefit formula, eligibility, and taxation of benefits. With a closed-group debt of $11.9 trillion, the program will be reformed in some way in the future. Thus, for a parallel comparison, the cost of a guarantee would have to be made explicit for both a prepaid program and the continuation of the status quo. Additionally, the existence of a risk-free rate is itself guaranteed by taxpayers.

[14] Admittedly there would be redistribution issues that would have to be addressed given that higher income workers' annuities would more than offset their scheduled benefits, but our purpose here is to merely illustrate the timing of the aggregate burden of a transition to fully prepaid accounts.

REFERENCES

Board of Trustees of the Federal Old-Age and Survivors Insurance and Disability Insurance Trust Funds. 2003. *2003 Annual Report of the Board of Trustees of the Federal Old-Age and Survivors Insurance and Disability Insurance Trust Funds*, Washington, DC, March.

Feldstein, Martin S. 1995. "Would Privatizing Social Security Raise Economic Welfare?" NBER Working Paper 5281. Cambridge, MA: National Bureau of Economic Research, September.

Feldstein, Martin S., and Andrew A. Samwick. 1997. "The Economics of Prefunding Social Security and Medicare Benefits." NBER Working Paper 6055. Cambridge, MA: National Bureau of Economic Research, June.

Feldstein, Martin S., Elena Ranguelova, and Andrew A. Samwick. 1999. "The Transition to Investment-Based Social Security When Portfolio Returns and Capital Profitability Are Uncertain." NBER Working Paper 7016. Cambridge, MA: National Bureau of Economic Research, March.

Friedman, Milton, and Rose D. Friedman. 1980. *Free to Choose: A Personal Statement*, New York: Harcourt Brace Jovanovich.

Garrett, Daniel M. 1995. "The Effects of Differential Mortality Rates on the Progressivity of Social Security." *Economic Inquiry* 33 (July): 457–75.

Geanakoplos, John, O. S. Mitchell, and S. P. Zeldes. 1998. "Would a Privatized Social Security System Really Pay a Higher Rate of Return?" Pension Research Council Working Paper 98-6. Philadelphia: University of Pennsylvania, June.

Kotlikoff, Laurence J. 1998. "Simulating the Privatization of Social Security in General Equilibrium." In *Privatizing Social Security*, ed. Martin Feldstein. Chicago: University of Chicago Press.

Kotlikoff, Laurence J., and Jeffrey Sachs. 1998. "It Is High Time to Privatize." Federal Reserve Bank of St. Louis *Review*, Vol. 80, No. 2, March/April.

Kotlikoff, Laurence J., Kent A. Smetters, and Jan Walliser. 1998. "Social Security: Privatization and Progressivity." *American Economic Review* 88 (May): 137–41.

Liu, Liqun, and Andrew J. Rettenmaier. 2003. "Social Security Outcomes by Racial and Education Groups." *Southern Economic Journal* 69: 842–64.

Liu, Liqun, Andrew J. Rettenmaier, and Thomas R. Saving. 2000. "Constraints on Big Bang Solutions: The Case of Intergenerational Transfers." *Journal of Institutional and Theoretical Economics* 156 (March): 270–91.

———. 2003. "Measures of Federal Liabilities." Unpublished manuscript.

Mariger, Randall P. 1997. "Social Security Privatization: What It Can and Cannot Accomplish." Manuscript, Board of Governors of the Federal Reserve System, June.

Murphy, Kevin M., and Finis Welch. 1998. "Perspectives on the Social Security Crisis and Proposed Solutions." *American Economic Review* 88 (May): 142–50.

President's Commission to Strengthen Social Security. 2001. *Strengthening Social Security and Creating Personal Wealth for All Americans*, Dec. 21.

Samuelson, Paul A. 1967. "Social Security." *Newsweek*, Feb. 13, 88.

Session 4

*Commerce, Culture, and Diversity:
Some Friedmanesque Themes in Trade and the Arts*
Tyler Cowen

*Milton and Rose Friedman's "Free to Choose"
and Its Impact in the Global Movement Toward
Free Market Policy: 1979–2003*
Peter J. Boettke

Free to Choose in China
Gregory C. Chow

Commerce, Culture, and Diversity: Some Friedmanesque Themes in Trade and the Arts

Tyler Cowen

Freedom and prosperity stand as the two central themes of Milton Friedman's political writings. Rather than offering either a "rights/liberty" defense of capitalism or a "utilitarian" defense, Friedman sought to identify the numerous cases where the two motivations coincide. Much of his analysis of markets has attempted to show this broader consilience or compatibility of values.

I wish to pursue this theme of consilience in more detail. Specifically, what other values might capitalism bring? Might markets and trade be the best means of encouraging the creative arts?

Friedman's works contain few explicit references to the creative arts (I survey some examples further below). Yet when it comes to the arts, in comparative terms capitalism and its accompanying wealth and liberties do the best job in delivering the goods. Friedman recognized this fact, albeit briefly and in passing. In his essay on why Jews are skeptical about capitalism, he wrote eloquently: "If, like me, you regard competitive capitalism as the economic system that is most favorable to individual freedom, to creative accomplishments in technology and the arts, and to the widest possible opportunities for the ordinary man, then you will regard Sombart's assignment to the Jews of a key role in the development of capitalism as high praise."[1]

I will follow this theme of capitalism and the arts, with special reference to Friedman's works. Friedman himself never made an *aesthetic* case for a market economy, above and beyond his liberty and utilitarian arguments. Such an aesthetic case remains underexplored, but Friedman's writings offer some useful tips for broadening our understanding of trade into the cultural dimension.

Similarly, a look at the arts can strengthen Friedman's overall case for free trade and globalization. Friedman has led a long crusade on behalf of economic globalization, from a classical liberal and cosmopolitan point of view. He has been a leading and vocal supporter of free trade, investment, and migration. I wish to consider how Friedman's analysis of these policies might contribute to our understanding of the aesthetics of capitalism.[2]

Friedman would be the first to admit that the principles governing international trade are the same as the principles governing domestic trade (for example, tariffs across countries make no more sense than tariffs across different counties, states, or cities). We cannot understand the benefits of international trade without a notion of how domestic trade works. So this brief investigation of trade, liberty, and the arts will consider both domestic and international factors, focusing on the Friedmanesque theme of how capitalism has supported diversity and creativity.

I will break the major topics into three parts. First, I will look at the general connection between commerce and diversity. Second, I will consider how governments, and government regulations in particular, can harm diversity. Third, I will look at international trade more generally and its effects on cultural diversity. Throughout I will keep an eye on Friedman's own writings of relevance, whether directly or indirectly, for these topics.

COMMERCE SUPPORTS DIVERSITY

A wealthy, free, and commercial society offers a diverse plenitude of cultural creations. America in the twentieth century, for instance, developed cinema, jazz, the skyscraper, rhythm and blues, rock and roll, abstract expressionism, pop art, science fiction, a long array of "highbrow" writers from Faulkner to Philip Roth, and numerous television shows, to name just a few items off a lengthy and varied menu. It is impossible for a single individual to come close to knowing all of the notable cultural achievements of the twentieth century.

Furthermore, cultural prices have become remarkably cheap. Owning original paintings by first-rate masters remains expensive, but when it comes to books, music, movies, and museum visits, we can usually get masterpieces, or just sheer entertainment, for less than $20 per experience. If we consider radio, libraries, and borrowing from friends, the dollar price is often zero. Even at Sotheby's, most of the items auctioned are worth less than $5,000. Today's upper middle class can now own beautiful artworks and collectibles. Beauty, fashion, design, and aesthetic inspiration—as expressed in concrete material forms—suffuse our lives as never before, largely due to capitalism.

Commercial society also does a notable job of preserving the cultural creations of the past. More people know Shakespeare or Mozart today than in past times, largely because private institutions repackage these creations for profitable sale, whether through compact discs, books, or movies. Mel Gibson and Martin Scorsese have made serious movies of the Gospels. The United States has experienced a museum boom for several decades, largely because wealthy private donors have shown a willingness to support these institutions. These art museums present a wide variety of styles and periods, not just American art or the popular arts. It is only wealthy societies that have the resources to take an interest in preserving their pasts or the pasts of other societies. Some of the best collections of Asian art can be found in the United States.

Increasing diversity has been the trend in virtually all areas. The number of musical genres available on compact disc, or in concert, has grown steadily. Book superstores have brought many different kinds of books to American cities and suburbs, not just bestsellers. Americans can now eat cuisines from around the world, not just Chinese and Italian food. My own metropolitan area (Washington, D.C., and northern Virginia) offers restaurants from such diverse locales as Bolivia, Afghanistan, El Salvador, Russia, Portugal, Peru, and Ethiopia, among many others, in addition to American regional fare.

Of course, we will not agree on which cultural innovations are the important ones, from a cultural point of view. The Beatles, John Updike, Andy Warhol, and *High Noon* all have their partisans and their detractors. But a market economy has an amazing ability to economize on consensus. The available variety is so great that people with many differing tastes can find strong favorites, without requiring that others follow the same path. This matching is one of the primary benefits of a market economy.

The Internet has rapidly become a major force for diversity. Individuals use the Internet to buy and research books (amazon.com or Powell's), to buy art (eBay), to follow their favorite musical group (various home pages), or to self-publish poetry. Blogs have become an alternative to mainstream journalism and opinion commentary. XM Satellite Radio offers one hundred stations for $10 a month; most of these stations are directed toward niche tastes, and many are commercial free. I grew up with only five or six TV channels, but now digital cable offers up to 500. Some critics charge that corporate conglomerates dominate cultural distribution, but the evidence indicates a clear move in the opposite direction, toward greater decentralization.

This phenomenal growth in diversity has occurred, for the most part, without significant government subsidy. Instead, American cultural institutions have funded themselves through fee for sale or donations. In recent years the budget of the National Endowment for the Arts (NEA) has ranged from $100 million to $115 million, and of course some of the funds are spent on staff and bureaucracy. For purposes of contrast, it cost $200 million to make the movie *Titanic*. State and local governments spend more, but the basic story remains one of private markets. Our most effective arts policy has been tax incentives for donations, which has kept choice and quality control in private hands.

Most for-profit creative enterprises get little or nothing from American governments, beyond enjoying the basic supply of public goods. Compact discs or Hollywood movies have to pass a market test. If we look at the nonprofits, such as the American symphony orchestras, 33 percent of their income comes from private donations and 16 percent from endowments and related sources. Concert income generates 42 percent of revenue, and direct government support provides only 6 percent of revenue. For nonprofit art institutions more generally, individual, corporate, and foundation donors make up about 45 percent of the budget. Twelve percent of their income comes from foundation grants alone, two and a half times more than the NEA and state arts councils combined.[3]

We can judge the aesthetic performance of a market economy by two major standards. The first standard, as we find in Friedman, is to ask whether the market satisfies the preferences of consumers. Here the answer is clearly yes. If anything, commentators criticize current cultural institutions for being *too* responsive to consumer demand.

The second standard, favored by many art lovers, is whether a market economy produces cultural masterpieces that will stand the test of time and last the ages. With respect to this question, it may be too early to judge what from the twentieth century will go down as a masterpiece, as noted in the above discussion of consensus. But wealthy societies from previous eras have a consistent record in being artistic leaders and producing masterpieces that stand a test of time. Renaissance Italy, for instance, used its wealth to fund notable paintings and sculptures; most of the relevant commissions were privately funded. The Dutch golden age of the seventeenth century relied almost exclusively upon private patronage, as did the French cultural blossoming of the nineteenth century. Mozart, Haydn, and Beethoven, contrary to some myths, all earned a good living in markets and managed to reach significant European audiences. The cultural peaks of the Chinese and Arabic worlds also coincide with their commercial successes. In similar fashion, we can expect the best creations of the twentieth century—whatever they turn out to be in the eyes of critics—to stand the test of time as well.[4]

Whenever the workings of a market economy are examined, we see evidence of an "anticapitalistic mentality." Many observers compare the plenitude of contemporary creations to the best of the past and find the modern world wanting. They forget that previous eras had their share of junk as well and that the best work needs time to rise to the top. Mozart was well-regarded in his lifetime, but he was not considered the greatest composer in Europe.

In other cases, many people, most of all intellectuals, object when apparently nonmeritorious individuals earn huge salaries. The same objections surface in the cultural realm. Madonna earns hundreds of millions, whereas a first-rate opera singer might pull in only $50,000 a year or perhaps cannot earn a living from singing at all. The best response, well understood by Friedman, is the same. A system that permits such "inequities" will in fact generate the greatest number of opportunities for performers of virtually all kinds. Government-sponsored arts programs, *if* done well, can support some narrowly defined areas of cultural excellence (the former Soviet Union, for instance, produced wonderful romantic pianists, not to mention chess players). But in terms of overall diversity, creativity, and satisfaction of consumer choice, the marketplace has by far the superior record.

Friedman on Cultural Diversity

The links between capitalism, wealth, competition, and diversity are a consistent theme in Friedman's writings. For instance, Friedman's analysis of school vouchers has involved numerous implicit references to culture. He points out, for instance, that school vouchers would allow various minorities to educate their children as they wish. Imagine that market and government schools competed on equal fiscal terms. We can imagine many more schools for liberals, conservatives, computer nerds, born-again Christians, strict disciplinarians, Catholics, Montessori advocates, and so on, not to mention various forms of home schooling. We would not get schooling of a single kind, but instead the schooling market would develop in many diverse directions.

Friedman (1975 [1972], 269) once noted: "Parents could escape a homogenized school system by sending their children to private schools." His vision of free market schooling always has involved many competing schools, not a single dominant school.

Friedman well understands the incentive for market entrepreneurs to lower costs of all kinds, including fixed costs, through innovation. Note that sufficiently high fixed costs would limit the ability of a voucher program to serve diversity. Assume, for instance, that in a market setting, the Baltimore area could support only a single high school. Competition would be weak, and parents could not send children to the "school of their choice." That single profit-maximizing high school would instead attempt to serve some weighted average of market demand (more technically, the school would focus on inframarginal consumers, whose decisions to pay hang most in the balance). We would not escape from schooling aimed at the least common denominator, so to speak, as we find in so many public schools today.

But the "natural monopoly" vision of homogenized market schooling does not square with the facts. Friedman has stressed repeatedly how rarely we find natural monopoly in a true market setting. He once remarked that only the New York Stock Exchange fit his notion of a truly market natural monopoly.[5] At the time, the NYSE, which of course can limit the number of seats it sells, held a dominant role in the trading of stocks. Even this case, however, has not held up as a true market monopoly over time. NASDAQ has risen in relative position over the last few decades, even with the burst of the dot-com bubble. Most major shares are traded on foreign exchanges as well. Off-floor trading and electronic trading continue to flourish. The NYSE now runs the risk of becoming a dinosaur, rather than a natural monopoly, and this was the best example of natural market monopoly that Friedman was able to find.

Friedman's analysis of television reflected this same theme of how commerce brings diversity. In 1969 he wrote a *Newsweek* column called "How to Free TV." He noted that the three major networks all broadcast the same point of view. He believed that they sought to present the news fairly, but that a low-quality and homogenized product was inevitable, given the incentives in place. The number of channels was limited by law, and restrictions on cable forced broadcasters to look to advertisements as their sole source of revenue. For purposes of comparison, imagine that only three printing presses were allowed, and they had to fund themselves through advertising revenue, rather than pricing the books in a market. What kind of books could we expect under such a system?

Friedman's recipe for improved television is unsurprising—increased reliance on market mechanisms. He (1975 [1969], 238) wrote with his characteristic bluntness: "This narrow range of views [on TV] has its origins in two related features of TV: first, the requirement of a government license in order to operate a TV station; second, the effective stifling of pay-TV for well over a decade by the Federal Communications Commission under the pressure and influence of the networks." He (239) notes that the FCC has been captured by the major networks, and has become an instrument "to preserve monopoly and prevent competition." He (238–39) describes the status quo as offering "deadening uniformity; limited choice; low-cost, low-quality programs." Many cultural critics today ascribe these features to the globalization of culture; Friedman, in his typically prescient fashion, laid them at the door of government intervention.[6]

GOVERNMENT REGULATIONS HARM CULTURAL DIVERSITY

In relative terms, and with the exception of television, America's cultural sectors have escaped many costs of regulation. The first amendment, with its guarantees of free speech, has made it harder for government to control many cultural outputs.

That being said, regulatory costs still have a significant and negative impact on cultural industries and on creativity more generally. I have never seen a truly reliable estimate of the costs of regulation (they are too high and too diffuse to count accurately), but some sources claim losses of up to $700 billion or $800 billion a year (Crews 2003), which comes out to several thousand dollars per American family. This decrease in wealth limits our opportunities to spend discretionary income, which of course hurts the cultural sectors disproportionately.[7]

Regulations also raise the costs of producing culture and lower diversity. To pursue another Friedmanesque theme, most government regulation increases the fixed costs of running businesses of all kinds, including cultural industries.

In the language of the economist, fixed costs are the bane of product diversity. A fixed cost means that some expenditure or investment must be made before an individual or firm can enter the relevant industry. These costs lower the number of competitors and impose something akin to a minimum scale requirement on the industry. Note, of course, that we do not see these costs with our eyes, since the relevant firms and products never come into existence in the first place. Frédéric Bastiat's distinction between "the seen" and "the unseen" is of course a longtime favorite idea of Friedman's.

Regulations commonly increase fixed costs and lower product diversity. Businesses above a certain size, or sometimes of any size at all, must meet various government regulations to stay in business. They must satisfy OSHA requirements, familiarize themselves with their complex legal liabilities, verify the immigration status of their employees, file complex tax returns, face the prospect of daunting environmental regulations, and fill out numerous forms of bureaucratic compliance. And this is only a partial list of the burden of regulation. Cultural firms, of course, bear these same costs.

Regulations, by raising costs, limit the number of firms that enter the market and thus limit diversity. Furthermore, large businesses can handle the regulatory burden far better than small firms can. (Indeed, we often see large firms pushing for additional regulation, for this reason.) Big firms can hire lawyers, tax accountants, and regulatory specialists. Small firms have less capital and less ability to manage these kinds of employees.

This penalty on small firms has special implications for culture. It is common in cultural sectors that small firms drive innovation. If we look at the market for popular music, for instance, smaller firms have initiated many of the breakthroughs. The Beatles were signed by a small firm (Vee-Jay) before the big record companies would market their songs in the United States; the majors apparently considered them to be too risky. Sun Records of Memphis gave a start to Johnny Cash, Jerry Lee Lewis, and Elvis Presley, among other notables. Berry Gordy of Detroit drove the Motown operation. Rap and heavy metal were rejected by the majors and picked up only after they succeeded with smaller, independent labels.

We find a similar pattern of small-firm innovation in cinema. Spike Lee, Martin Scorsese, the Coen brothers, Francis Coppola, Jonathan Demme, David Lynch, John Sayles, and many other prominent directors got their start with "microbudget" films, made with independent film companies or with their own capital. Only later, once they had proven their quality, did they have subsequent opportunities to make more expensive movies. In similar fashion, painters and sculptors use smaller galleries as a stepping stone to the major galleries and museums.

The costs of regulation apply to the independent artist as well, who faces legal, tax, and regulatory burdens. An aspiring independent musician, for instance, faces a bewildering array of tax schedules and filing options. Copyright law is hardly transparent (subsequent rights, for instance, may depend on whether the initial creation was "work for hire," a distinction understood by few musicians). Entire books are written to give artists and musicians guidance in these legal issues. While these tomes are not above the head of a lawyer or Ph.D. economist, I suspect they bewilder most musicians, many of whom are focused on their art. If an individual is truly bent on investing his or her energies in the creative process, simple and understandable laws offer significant benefits.

We therefore can think of government regulation as limiting the diversity of our culture. By shutting down small firms and single-artist operations, or stopping them from getting started in the first place, government regulation limits the number and kind of cultural delights.

FREE INTERNATIONAL TRADE BENEFITS CULTURAL DIVERSITY

Just as domestic commerce brings diversity and cultural riches, so does international trade. Indeed, both economic theory and the cosmopolitan view suggest that there is nothing special about the international case. Markets should perform well across borders, just as they work well within borders.

Critics, both on the left and right wing, commonly charge that we are headed toward a homogeneous global culture of the "least common denominator." McDonalds, Reebok, and Ricky Martin are examples of the supposed sins of global trade in culture. Perhaps no issue today drives greater hostility to markets, globalization, and free trade across nations.[8]

That being said, today's intellectual elites, including the critics of globalization, rely on globalization like never before. Many American academics, for instance, will shop for French cheeses, buy Japanese automobiles and stereo systems, vacation abroad, use the Internet to write friends in foreign countries, and rent foreign films, all while complaining about the cultural impact of globalization.

A look at the facts shows globalization to be more of a cultural hero than villain. For instance, many non-Western literatures were making few advances until the Western printing press and bookstore came along. Excellent movies are now made around the world (Taiwan, Iran, Hong Kong, and India are some favorite cinematic sources of mine), but the core technologies are Western. Acrylic paints, a product of largely Western technologies, are now used by artists around the world, as is the metal carving knife.

It is difficult to find a cultural product or creation that is not based on trade and cosmopolitan principles. Consider the book. The Chinese invented paper, the Western alphabet comes from Phoenician culture, page numbers are an Arabic and Indian innovation, and the history of printing runs through Germans (Gutenberg) as well as Chinese and Koreans. Friedman (1980, 11–13, with Rose Friedman) uses the "I, Pencil" example to illustrate the international division of labor; the history of the book shows the generality of this example.

The histories of specific arts illustrate similar themes about the benefits of trade and division of labor. For instance, the so-called golden age of Persian carpets came largely in the seventeenth century. At this time Persia was a stable region with an extensive network of trading connections. Most high-quality Persian carpets were made for export, not only to Europe but also to the Arabic elites of the Ottoman Empire and to India. Without foreign buyers, and the possibility of trade, the Persian carpet tradition could not have flourished. It is no accident that we see so many Persian carpets in the paintings of Van Dyck, Vermeer, Rubens, and others.

Persian carpet making dried up in the eighteenth century and on a large scale came to a virtual halt. Persia lost its political stability, and international trade networks collapsed. Persian property rights no longer were stable. The large-scale carpet factories no longer were profitable, and most of them were closed down, although tribal carpets continued to be made.

Persian carpet making was revived only in the nineteenth century, largely because of contact with the wealthier West. Europeans and North Americans suddenly had great concentrations of industrial wealth and were looking to buy new fineries. Carpet marketing spread to the West quickly, with the aid of high-quality department stores, such as Liberty in London and W. J. Sloane in New York. The Persian workshops restarted, often with the aid of foreign capital; many of them were now owned and run by British and German firms. Production was geared up quickly, and a second golden age of Persian carpet making was under way. Many masterpieces date from this era, and the latter nineteenth century boom produced many more high-quality carpets than the Persians had managed before, largely because of trade with wealthy buyers from other countries.[9]

This story of free trade and creativity runs throughout the history of culture. Claude Monet had little success marketing his paintings to the government-run Salon in Paris in the late nineteenth century. His style and colors were considered to be too radical and too unpleasant. Monet had greater success selling to wealthy North Americans, who were not bound by prevailing French artistic conventions. His haystack paintings proved particularly popular in this country, which is one reason why they appear so frequently in American art museums.

The Monet example illustrates a broader (but sometimes neglected) benefit of international trade. The common arguments for trade cite the benefits of drawing on producers from other countries. But trade also mobilizes the benefits of the *consumers* from other countries. Consumers hold embedded knowledge. Their purchases can induce suppliers to elevate quality, help suppliers pursue careers of greater pleasure (for example, art), and help generate the artistic heritage of mankind. The greater the diversity of consumers to draw on, the better markets will perform these tasks. International trade, of course, maximizes the diversity of consumers to the greatest extent possible.

Nor is the case of Monet a unique or outdated example. To provide a more modern example, the music of Jamaica has relied on foreign buyers since the late 1960s. Since that time, North American and UK buyers have accounted for more revenue than Jamaican buyers. The growth of the market has allowed Jamaican music to become very innovative, very popular around the world, and also very diverse. We think commonly of reggae, but in fact Jamaica has supported many kinds of music, including dancehall, lovers' rock, ska, mento, ragga, and dub.

We might think of Jamaica as a prime candidate for the model of cultural imperialism. After all, it is very close to the United States, a former British colony, English-speaking (albeit with dialects), very small, and relatively poor. As late as 1950, Jamaica had no recording industry of its own. We might think that Jamaican music would simply be overwhelmed by American music, but this has not been the case. Jamaican music borrowed from American (and British) music without being dominated by it. Jamaican popular music borrowed from American rhythm and blues, heard over radio broadcasts from New Orleans, but rapidly pursued its own course. Since this time, American and British music has arguably borrowed as much from Jamaican music as the other way around. Paul Simon, Paul McCartney, Blondie, and the Clash have all looked to Jamaica for musical inspiration. Electronic music, such as techno, jungle, and rave, took a big initial cue from the Jamaican dub style. Jamaican artists Shaggy and Sean Paul have topped the American charts in recent years.

All of these examples represent a more general historical pattern. Eras with growing international trade tend to be creative and diverse; eras with shrinking international trade tend to exhibit cultural decline. For instance, the period between 1800 and World War I brought an unprecedented boost to globalization. The steamship, the railroad, and the motorcar, embedded within a broadly classical liberal European order, supported international trade, investment, and migration. The nineteenth century in turn was an extremely creative, diverse, and culturally fruitful time.

In contrast, the most prominent period of cultural decline for the West coincides with falling trade relations. The Dark Ages that followed the collapse of the Roman Empire brought a massive contraction of foreign trade and investment. Trade routes fell into disuse, cities fell, and nobles retreated to heavily guarded country estates, giving rise to feudalism. Architecture, writing, reading, and the visual arts all declined during this period. The buildings of antiquity fell into disrepair or were looted and destroyed. Greek bronzes were melted down for their metal, and many books and plays from the antique world were lost.

A society with retrogressing trade relations will find it harder to innovate and harder to preserve the best of its past.

The critics of globalization often confuse differing kinds of diversity. Trade does often decrease diversity *across* societies. That is, different places become more alike. But these societies become more alike by offering more choice across the board. Today it is possible to buy Milton Friedman's writings in Germany, France, China, Russia, and Mexico, among many other countries. But these societies have become more alike by offering more choice, a commonly diverse menu of options. So diversity *within* societies goes up.

Alternatively, it can be said that diversity for individuals goes up, even though diversity for collectives may fall. Individual Americans can now choose from more differing life paths, and from more differing cultural items, than ever before. It is this individual notion of diversity that is most important for economists, and most important for Friedman, who emphasizes consumer sovereignty at the individual level. Yet, at the same time, societies are more similar in the aggregate and crossing a border is less of a shock than it used to be.

The Friedmans on Tourism and Globalization

Milton Friedman never outlines such a cultural vision for international trade, but I hope he would welcome the overall tenor of these remarks. The Friedmans' memoir, however, offers some briefly skeptical remarks about cultural globalization. They write:

> The character of Bali had changed since we visited it a third of a century earlier. Tourism had overwhelmed it. We had brought back beautiful carvings from Bali on our 1963 visit. This time, everything seemed to be mass-produced. Hugo took us to the best current carvers and craftsmen, but we were unable to find any small-scale carvings that seemed to us to match in quality the ones we already had. No doubt they exist, but they account for a much smaller fraction of the market. (Friedman and Friedman 1998, 581)

Consistent with the analysis from above, I believe that the effects of tourism on Bali are more positive than this passage would suggest. First, the island of Bali is very small and relatively poor. Tourism, directly or indirectly, accounts for most of the economic activity. Without tourism, Bali probably would be depopulated and run down. It is easy to see what tourism has ruined, but without tourism the island's culture would not have been preserved in the first place.

Second, tourism has a long history of supporting native Balinese art forms. Sculpting, naïve painting, Balinese dance, and Balinese music all have owed significant debts to tourist demands and foreign influences. Dances are preserved to market to tourists, and some of these dances draw upon foreign inspirations.

Perhaps the most famous Balinese dance is the Kecak, where dozens of Balinese sing the rhythmic vocal of the "monkey chant" while waving their upper body and arms. Walter Spies, a German artist, choreographed the Kecak in 1932 for a German film (*The Island of Demons*). Even if Balinese carving has declined in quality more recently, we must evaluate tourism in terms of this overall picture.

Third, even the Friedmans, obviously two authors sympathetic to the market, may have confused the issue of average quality with the question of whether consumers get what they want. The Friedmans, for instance, probably would not write that the automobile industry has been "overwhelmed" by the cheap demands of the ordinary public. Indeed, toward the end of the passage the Friedmans note that high-quality Balinese carvings still probably exist. The most likely scenario is the Smithian story of "the division of labor is limited by the extent of the market." Now that the demand for Balinese carvings has grown, we would expect to find carvings of many different kinds, and of many different qualities. There will be more low-quality carvings, but not necessarily at the expense of high-quality carvings. The casual tourist may find it difficult to sort through the larger market, but the same could be said for just about any other market. The more choice in that market, the more bewildering that market can be to the uninformed.

Refer to the distinction between diversity for individual choice and the more collectivist question of how much different geographic regions resemble each other. Bali may have become less diverse in the sense of offering commodities, namely cheap carvings, that the richer countries offer as well. At the same time, diversity within Bali has gone up, as it is now possible to buy either very good or very cheap carvings within Bali. People in Bali, be they tourists or natives, have a richer menu of choice.

In contrast to this case, Friedman was more optimistic about another instance of cross-cultural clash—the West Bank in the Middle East. Friedman visited the West Bank in 1969 and wrote the following for his *Newsweek* column:

> Much to my surprise, there was almost no sign of a military presence.... I had no feeling whatsoever of being in occupied territory.... This wise policy [of the Israelis] involved almost literal laissez-faire in the economic sphere.... To a casual observer, the area appears to be prospering. (Friedman 1975 [1969], 298–99)

Why be so optimistic about the West Bank and so relatively pessimistic about Bali? We can only speculate about the answer. In part, many commentators did not foresee the 1973 Yom Kippur War, which made it harder for Israel to pursue liberal policies. In part the Friedmans' earlier visit to Bali may have led them to expect the "idyllic paradise" to continue, whereas he clearly expresses surprise at seeing the West Bank as anything but an armed camp ("Much to my surprise, there was almost no sign of a military presence," 298).

So many of our evaluations, including those of a market economy, are relative to our expectations, and the Friedmans, however astute their observations, are no exception in this regard.

CONCLUDING REMARKS

Milton Friedman already has gone down as one of the most important thinkers and social scientists of the twentieth century. In addition to his numerous famous contributions, his works hold significant lessons for anyone approaching an analysis of culture, diversity, and aesthetics.

NOTES

The author wishes to thank Peter Boettke, Bryan Caplan, and Alex Tabarrok for useful comments.

[1] See Friedman (1987, 53) [1984]. This interesting and underrated piece represents one of Friedman's rare forays into cultural analysis. He sought to explain why, in the course of European history, so many Jews had become socialists or expressed opposition to a market economy.

[2] The legacy does not end there. Friedman was instrumental behind the development of currency futures at the Chicago Merc in the 1970s. These hedging and risk management instruments have increased the volume of international trade and investment and gave Friedman a permanent place in history as a practitioner and not just an economist.

[3] See Johnson (1997, 9); the data refer to 1995. On foundations, see Dowie (2001, 169).

[4] My *In Praise of Commercial Culture* (Cowen 1998) discusses these claims in more detail.

[5] I recall hearing this comment on his *Free to Choose* TV series.

[6] It is an interesting theoretical question why private monopoly might be expected to damage product diversity. After all, a single firm can supply many different kinds of products, and monopolies still have incentives to innovate. Most likely, the presence of only a single firm would limit the number of sources of new ideas and limit cross-firm learning externalities. But in some sectors monopoly may encourage rather than discourage innovation, a claim that dates at least as far back as Joseph Schumpeter. For a survey of the relevant literature, see Kamien and Schwartz (1982).

[7] Oddly, the income tax has a partially positive effect on culture. Of course, the negative income effect lowers the demand for culture. But the substitution effect encourages additional interest in fun, lower-paying jobs, which probably includes many cultural sectors. And a given creator will be more likely to produce as he sees fit, rather than maximize (taxable) profits by meeting market demand.

[8] France, Spain, Canada, Brazil, and South Korea are among the nations that practice cultural protectionism. For a critique of cultural globalization, see Barber (1995).

[9] See Cowen (2002) on this whole episode.

REFERENCES

Barber, Benjamin R. 1995. *Jihad vs. McWorld.* New York: Times Books.

Cowen, Tyler. 1998. *In Praise of Commercial Culture.* Cambridge, MA: Harvard University Press.

———. 2002. *Creative Destruction: How Globalization Is Changing the World's Cultures.* Princeton, NJ: Princeton University Press.

Crews, Clyde Wayne, Jr. 2003. *Ten Thousand Commandments: An Annual Snapshot of the Federal Regulatory State*, 2003 edition. Washington, DC: Cato Institute.

Dowie, Mark. 2001. *American Foundations: An Investigative History.* Cambridge, MA: MIT Press.

Friedman, Milton, with the assistance of Rose Friedman. 1962. *Capitalism and Freedom.* Chicago: University of Chicago Press.

Friedman, Milton. 1975 [1969]. "How to Free TV." In *There's No Such Thing as a Free Lunch: Essays on Public Policy*, 237–40. LaSalle, IL: Open Court Press.

———. 1975 [1972]. "Homogenized Schools." In *There's No Such Thing as a Free Lunch: Essays on Public Policy*, 268–70. LaSalle, IL: Open Court Press.

———. 1975 [1969]. "Invisible Occupation." In *There's No Such Thing as a Free Lunch: Essays on Public Policy*, 298–99. LaSalle, IL: Open Court Press.

———. 1987 [1984]. "Capitalism and the Jews: Confronting a Paradox." In *The Essence of Friedman*, ed. Kurt R. Leube, 43–56. Stanford, CA: Hoover Institution Press.

———. 1987. *The Essence of Friedman*, ed. Kurt R. Leube. Stanford, CA: Hoover Institution Press.

Friedman, Milton, and Rose D. Friedman. 1980. *Free to Choose: A Personal Statement.* New York: Harcourt Brace Jovanovich.

———. 1998. *Two Lucky People: Memoirs.* Chicago: University of Chicago Press.

Johnson, Arthur T. 1997. "Symphony Orchestras and Local Governments." Working paper, Maryland Institute for Policy Analysis and Research, University of Maryland, Baltimore County.

Kamien, Morton I., and Nancy Lou Schwartz. 1982. *Market Structure and Innovation.* Cambridge, MA: Cambridge University Press.

Milton and Rose Friedman's "Free to Choose" and Its Impact in the Global Movement Toward Free Market Policy: 1979–2003

Peter J. Boettke

In 1964, Lyndon Johnson defeated Barry Goldwater for the presidency of the United States by the overwhelming margin of 61 percent of the popular vote to 38 percent, and in terms of states won, the figure was forty-four to six. Barry Goldwater ran a campaign calling for less government and freer markets, and the population said no to him and yes to Lyndon Johnson's big government programs of the 1960s, for example, the War on Poverty. However, in the 1980 election, Ronald Reagan was able to defeat the incumbent president, Jimmy Carter, with 51 percent of the popular vote to 41 percent, and in terms of states, forty-four states to six states, running on essentially a similar platform to Goldwater's.

Obviously, something had drastically changed in that intervening sixteen years in the United States. For sure, a good part of that was the failure of the welfare/warfare state in the 1960s and 1970s. The U.S. economy in the 1970s was suffering from declining productivity, growing public debt, and inflation. The declining stature of the United States as *the* economic leader in the world was matched by a declining stature as a military superpower—as the frustrations of Vietnam fed into the failed policies in the Middle East, most obviously brought home by the Iranian hostage crisis in 1979. Reagan ran on a platform to reverse all of that, and in so doing he captured the imaginations of many. In particular, his rhetoric of uncompromising adherence to free market economics signaled a change in political rhetoric and public opinion.

Since the beginning of the progressive era, laissez-faire economics had been on the run from intellectuals and politicians and, since the Great Depression, the general public. There were, of course, lone wolf voices bucking this trend all along: Ludwig Mises, F. A. Hayek, Henry Hazlitt, and Ayn Rand being perhaps the most prominent in the 1930s, 1940s, and 1950s. In the 1940s, a superstar economist emerged to add his voice to these lone wolves and challenge the Keynesian hegemony in the economics profession and the conventional wisdom in the court of public opinion—Milton Friedman. Friedman's accomplishments as an economist, and as the premier public intellectual in the second half of the twentieth century for economic liberalism, are well known, so that is not what I am going to emphasize here. Instead, I want to focus on how his work in conveying the basic principles of economic liberalism changed public attitudes in the United States and abroad among the political elite, the intelligentsia, and the educated public, and, in particular, how that success in changing the climate of public opinion in the West in turn represented a beacon of hope to those in East and Central Europe and the former Soviet Union during the years leading up to the collapse of communism in 1989 and 1991.

There are many hypotheses about why the communist system collapsed in the late 1980s and early 1990s. One is that a Polish pope gave legitimacy to the Church behind the Iron Curtain and the unrest with Solidarity discredited the Workers' State in Poland. Once the Polish communist regime collapsed, the others followed. Another hypothesis is that Ronald Reagan's decision to up the military stakes highlighted the technological gap between the economic systems and toppled the system. Still another hypothesis is that a generation of political leaders from within the communist system that came of age during 1956 (the "thaw generation") and knew firsthand of Stalin's crimes against humanity had decided that this was no way for a civilized people to live. I cannot do justice to all these competing hypotheses here, but I want to suggest an alternative one and provide evidence of its plausibility—namely, that the economic failures of the real-existing communist system in East and Central Europe made sense only in light of the ideas of economic liberalism.[1] And in the 1980s, no one had stated those ideas more plainly and concisely than Milton and Rose Friedman in *Free to Choose*.[2]

FROM *CAPITALISM AND FREEDOM* TO *FREE TO CHOOSE*

One way to measure the impact of Milton Friedman's ideas is to simply compare and contrast the reception of *Capitalism and Freedom* at the time of its publication in 1962 with that of *Free to Choose* in 1980. As the Friedmans inform us in their memoirs, the intellectual climate of opinion at the time of *Capitalism and Freedom* was, to put it mildly, hostile (Friedman and Friedman 1998, 339). Milton Friedman states in the preface to the 1982 edition of *Capitalism and Freedom* that when it was first published in 1962,

> its views were so far out of the mainstream that it was not reviewed by any major national publication—not by the *New York Times* or the *Herald Tribune* (then still being published in New York) or the *Chicago Tribune* or by *Time* or *Newsweek* or even the *Saturday Review*—though it was reviewed by the London *Economist* and by the major professional journals. And this for a book directed at the general public, written by a professor from a major

U.S. university, and destined to sell more than 400,000 copies in the next eighteen years. It is inconceivable that such a publication by an economist of comparable professional standing but favorable to the welfare state or socialism or communism would have received a similar silent treatment. (Friedman 1982, vi)

The publication of *Free to Choose* would provide the exact opposite experience for the Friedmans.[3] The book sold 400,000 copies in hardcover and as a mass market paperback has sold millions of copies and been translated into over a dozen languages.[4] Perhaps an even better measure, though harder to put a precise number to, is how proposals first discussed in *Capitalism and Freedom* (and considered too radical for respectable conversation) have now become commonplace: monetary rules versus discretionary policy, private certification on the market rather than government licensure, school vouchers and competition in education versus government monopoly, and the flat tax versus a progressive income tax are but a few examples of how Friedman was the trailblazer for creative applications of market thinking to areas of public policy.

In *Capitalism and Freedom*, Friedman sought to establish an argument about the interconnectedness of economic and political freedom. It was possible, he argued, to have limited political freedom while adopting policies of economic freedom, but it was impossible to eliminate economic freedom without also infringing on the political freedoms of individuals. Moreover, economic freedom would put pressures on the political system to open up. In contrast to the popular position among intellectuals that political and economic freedom could be separated neatly, Friedman put forth the following historical challenge:

> Historical experience speaks with a single voice on the relation between political freedom and a free market. I know of no example in time or place of a society that has been marked by a large measure of political freedom, and that has not also used something comparable to a free market to organize the bulk of economic activity. (Friedman 1982, 9)

While *Capitalism and Freedom* is in many ways a more philosophical and foundational book than *Free to Choose*, the basic teachings of economic liberalism are conveyed even more forcefully and the applications more persuasive than in the earlier book. Moreover, *Free to Choose* is more explicit in its use of ideas such as the informational role of prices, the spontaneous order of the market system, and the interest group logic of political interference with the market. These aspects of the theoretical foundations of liberalism were not emphasized in the early 1960s but emerged more explicitly with the development of public choice theory by James Buchanan and Gordon Tullock in *The Calculus of Consent* (1962) and the theory of spontaneous order in F. A. Hayek's work from *The Constitution of Liberty* (1960) to *Law, Legislation and Liberty* (1973, 1976, 1979).[5] In short, what the Friedmans argued in *Free to Choose* is that the power of the market system is its ability to mobilize the incentives of individuals to realize the gains from mutually beneficial exchange, and that the price system is an indispensable aid in this endeavor by discovering the relevant information and communicating it to the relevant actors within the system, who in turn utilize it efficiently in realizing their individual plans. On the other hand, the attempt by government to interfere in the market order results in perverse incentives, distorted information, and the catering of special interests that concentrates benefits on well-organized and well-informed interest groups and disperses the costs among the unorganized and ill-informed mass of voters.

The Friedmans summarize the functions of prices in a market economy as follows: "Prices perform three functions in organizing economic activity: first, they transmit information; second, they provide an incentive to adopt those methods of production that are least costly and thereby use available resources for the most highly valued purposes; third, they determine who gets how much of the product—the distribution of income" (Friedman and Friedman 1980, 6). The price system constitutes an intricate web of information and incentives. Attempts by government to substitute control for voluntary exchange often result in a failure to rectify whatever perceived problem was used to justify government action in the first place and, in fact, often exacerbate the problem by imposing costs on some parties and concentrating benefits on others. Freedom of trade fosters cooperation and harmony of interests among diverse parties. Controls lead to conflicts and special interest politics: "There is, as it were, an invisible hand in politics that operates in precisely the opposite direction of Adam Smith's invisible hand. Individuals who intend only to promote the *general interest* are led by the invisible political hand to promote a *special interest* that they had no intention to promote" (Friedman and Friedman 1980, 281).

Free to Choose leaves its reader with a clear message about the power of the market to harness individual initiative and knowledge of time and place, the importance of property rights and the rule of law in enabling individuals to realize the gains from exchange and to preserve our personal freedom, the failure of government policy to achieve the goals set, the vulnerability of government policy to opportunistic behavior by special interests, and the threat to human liberty that government intervention in the economy represents. While their message was directed primarily at an audience of U.S. readers, the Friedmans infused their work with a comparative analysis drawing on examples from Russia, India, China, and Hong Kong, among other places. The message learned through this comparative historical analysis is this:

> Wherever we find any large element of individual freedom, some measure of progress in the material comforts at the disposal of ordinary citizens, and widespread hope of further progress in the future, there we also find that

economic activity is organized mainly through the free market. Wherever the state undertakes to control in detail the economic activities of its citizens, wherever, that is, detailed central economic planning reigns, there ordinary citizens are in political fetters, have a low standard of living, and have little power to control their own destiny. (Friedman and Friedman 1980, 46)

It is this broad sweeping judgment that would serve as an inspiration and catalyst for dissident economists within the former Soviet Union to push for economic and political change in the late 1980s and early 1990s.

THE INDIRECT AND DIRECT INFLUENCE OF MILTON FRIEDMAN IN 1989 AND 1991

The indirect influence of Milton Friedman as the leading intellectual spokesman for economic liberalism stretches from China to Poland and all points north and south as well. Only Hayek's influence would compare.[6] This indirect influence is revealed anytime a modern economic reformer insists on the interrelationship between economic freedom and political freedom, on the necessity of private property and the freedom of contract, on the importance of rule-bound monetary and fiscal policy, on the perverse consequences of government regulation, and on the special interest groups that form the tyranny of the status quo. Friedman made it respectable for economists to argue in favor of free markets and offer market solutions to public policy questions. In his own attempts to provide market solutions in public debates, Friedman originated many of the ideas that defined not only the Thatcher and Reagan revolutions in the 1980s, but would define transition policies in Poland, the Czech Republic, and Russia in the 1990s. Many of these ideas were forged in Friedman's attempts in the 1960s, 1970s, and 1980s to address vexing policy problems in the United States, UK, India, Israel, Latin America, and China.

Far from just the figurehead of the political philosophy and political economy of classical liberalism that many of the reformers embraced in their rejection of the previous socialist system, Friedman was also an inspiration for many of the policy proposals adopted. The Friedmans did not dare in 1980 believe that communism would topple in a decade, but they also didn't rule that option out:

> [L]etting the genie of private initiative out of the bottle even to this limited extent [context is Yugoslavia in the 1970s] will give rise to political problems that, sooner or later, are likely to produce a reaction toward great authoritarianism. The opposite outcome, the collapse of communism and its replacement by a market system, seems far less likely, though as incurable optimists, we do not rule it out completely.[7] (Friedman and Friedman 1980, 49)

The problem with the real-existing systems in Soviet bloc countries could not be tackled coherently with minor reforms to the socialist system. The problem wasn't with this or that aspect of the system but the entire system.[8]

Milton Friedman (1984), in a pamphlet for the Centre for Research into Communist Economies, summed up his position on the problems of trying to introduce markets into a command economy by stating,

> I believe this way of putting it is upside down. The real question is how far one can go in introducing command elements into a market economy. I believe it would be literally impossible for any large-scale economy to be operated on a strictly command basis. Fundamentally, what enables a country such as China or the Soviet Union to function at all is the market elements that are either deliberately introduced or are inadvertently permitted to operate. When I speak of market elements being introduced into command economies such as China's and the Soviet Union's, I am not speaking of free markets; they are highly distorted markets. That is why those countries have such low standards of living; that is why they are so inefficient. (Friedman 1984, 8)

The power of Friedman's observations of the failure of the real-existing socialist economies of the Soviet bloc was not lost on those in charge of designing the reforms for those economies. Abel Aganbegyan, one of Gorbachev's main economic advisors during the 1980s, describes his meeting with Milton Friedman in San Francisco as follows:

> I was astonished by his fantastic faith in private property, a faith that excluded the possibility of any other kind of property ownership such as that which exists in the socialist countries. In Friedman's opinion, well-being can be reached only through private ownership of property, a free market and the existence of banks completely independent from the state and serving that free market....But if we move away from conceptual problems to the concrete theories advanced by Milton Friedman in his studies, we find that many of them can be of great use to us. In a number of cases Friedman points to examples of financial misjudgement by the state in increasing expenditure, printing excess money and so on. And while I do not accept his view that the socialist countries should transfer property into private ownership, I nevertheless listened with great interest to his explanations of the present inflation in China, which he had recently visited, and in other socialist countries. (Aganbegyan 1989, 52–53)

The Gorbachev reform team lacked the imagination to embrace private property and the market economy, and instead the inconsistency in their reform efforts led to the unraveling of the Soviet system.

When a group of young economists was tapped to form the new reform team at the end of the Gorbachev period and then into the Yeltsin years, Milton Friedman's influence was again repeatedly recognized. In her book *Sale of the Century*, Chrystia Freeland makes this stunning observation: "It was, of course, an absurd decision. Here was Gaidar, an ardent capitalist, a fan of F. A. Hayek and Milton Friedman, a man who thought the welfare state in Western Europe was far too large and would have voted for Ronald Reagan, shaping the economic ideology of the Communist Party of the Soviet Union (CPSU). It was like asking a crusading atheist to write a new catechism for the Vatican" (Freeland 2000, 29).[9]

Friedman's insights into the nature of real-existing communist economies were important for a variety of reasons that would later prove crucial during the transition period. It was the distorted market economy that failed in 1989 and 1991, with the social networks and political interconnections that had been formed under the incentives of that distorted system. The system led to disproportionate power to those in politically privileged positions, inefficiencies in production due to perverse incentives and the distorted signals of administered prices, and lacked any incentive for innovation, change, and progress. As Friedman would put it during a trip to China in 1988, "The problems of overcoming vested interests, of frustrating rent-seeking, apply to almost every attempt to change government policy, whether the change involves privatization, or eliminating military bases, or reducing subsidies, or anything else" (Friedman 1990, 94).

In order to defeat the vested interests and transition to a free market economy, Friedman counseled that reformers move quickly and decisively. The discussion is subtle because Friedman admits that "slow and steady" may outperform "one fell swoop" under certain conditions, and in particular may outperform with regard to issues of equity and political sustainability of the reforms. But ultimately, the arguments for economic efficiency, and the reality that gradualism enables vested interests to organize and fight against change, leads one to lean toward quick and decisive moves in economic policy. This does not mean that reformers should give little thought to the political sustainability of reforms. Instead, as Friedman highlights, there are a few basic ways to address the tyranny of the status quo in economic reform. One way, followed in the case of British Telecom, is to try to create stakeholders from the vested interests so they will see the benefits of privatization. As Friedman warns, the problem with this approach is if you end up simply substituting a private monopoly for a government monopoly, the politically connected will fight to maintain an effective barrier to entry in the respective industry.[10] To avoid this problem, Friedman himself advocated a free distribution of shares in the state enterprises and then allowing citizens to freely buy or sell the shares in an open market. And, finally, rather than fighting the existing monopoly head-on, the reformer could simply eliminate the government-enforced barrier to entry in the industry and allow the market to generate substitutes either through the entry of direct competitors or technological innovations that change the nature of the industry.[11]

While we have seen that the different economists in Russia acknowledged Friedman's influence in framing their discussions, the most successful economic reformers in Poland and the Czech Republic borrowed from Friedman more than a frame of reference. Poland's finance minister, Leszek Balcerowicz, has turned to Friedman to find practical ideas from monetary stabilization to privatization to the flat tax.[12] Jeffrey Sachs (1993, 87), Anders Aslund (2002, 256), and Marshall Goldman (2003, 196) all credit Friedman as the "godfather" of the voucher privatization proposals that circulated throughout transition economies. Perhaps the strongest endorsement of a direct influence of Milton Friedman guiding the transition experience comes from Vaclav Klaus, and thus it is worth quoting him at length from a speech he gave at the Prague School of Economics on the occasion of awarding an honorary doctorate to Milton Friedman on April 17, 1997:

> Reading and studying Milton Friedman's works helped me and many of us to understand economic reality, to understand economics, to understand its methodology, the role of the market in society, the role of the state in a free market economy, the role of money in the economy etc. Surely there were other influential authors but there was no one comparable in intellectual and human integrity, in firmness of stances and attitudes, in innovative boldness, in simplicity and clarity of exposition and in the scope and quality of important contributions both to economic theory and to the theory of public policy.
>
> Milton Friedman is, however, not only a theoretician in the very rigorous discipline of economic science. He is, at the same time, a true believer in the unrestricted market economy and I believe that his books *Capitalism and Freedom*, together with a more recent *Free to Choose*, opened the eyes of whole generations of not scholars but of ordinary citizens on all continents of this planet.
>
> All that helped us to understand the tenets of the old communist regime and its oppressive character and economic irrationality. With Milton Friedman's works as our background we had no dreams about the so-called third ways, about perestroika, about the reformability of communism. Milton Friedman helped us to interpret the actual communist economy not as a textbook command economy, based on directives going in the vertical direction from the central planning commission at the top to individual firms but as a very strange and truncated market economy with imperfect, but nevertheless dominant horizontal relations among economic agents at the microlevel. Milton Friedman knew that it was impossible to suppress human behaviour, the

spontaneity of exchange, implicit if not explicit prices, wide-spread bargaining etc. It was a very rare attitude at that time.

At the same time, the works of Milton Friedman helped us to understand the logic of the transformation of a communist country into a free society and a full-fledged market economy. Because of him, we had a clear vision where to go and a pragmatic strategy how to get there. We did not want to mastermind the whole process because it would not be possible and definitely not successful. We knew we had to trust free citizens to create the new world— with a moderate help from the above only.

Klaus's words touch on the themes we have emphasized as coming out of *Capitalism and Freedom* and *Free to Choose* with regard to the power of markets and the tyranny of controls, and clearly state how Friedman's ideas guided the construction of economic policy during the transition.

How have these policies fared? If you listen to the popular press and left-leaning academics, then you will hear about social disruptions and a general discrediting of market reforms in East and Central Europe and the former Soviet Union. But the popular rhetoric is often divorced considerably from the reality of the situation, especially as compared with merely a decade ago.[13] As Vladimir Dlouhy, the former minister of industry and trade of the Czech Republic has put it: "If someone would have told me in 1989 that by 2009 we would have a capitalist economy, the rule of law, a stable democracy, European integration, etc., etc., I would have told them they were crazy. When you look at the immediate past, you become a pessimist. When you look at a longer term horizon, the progress is breathtaking."[14] Of course, mistakes were made, and the corrupting influence of interest group politics is ever present. Intellectually, we now know that we must emphasize the necessity of certain key institutions in order for markets to work as effectively as we might hope—a point that is implicit in the Friedman analysis of the power of the market and tyranny of controls, but now must be made more explicit, as has been repeatedly stressed by James Buchanan (1997), Ronald Coase (1992), and Douglass North (1994). Moreover, culture and history no doubt represent a significant constraint on our ability to establish a successful market economy in the former communist economies, as recently stressed by Steve Pejovich (2003).

The impact of culture and history is not felt in terms of economic performance of market-oriented reforms *if* they were implemented.[15] Privatization and competition will lead to gains in productive efficiency and with that, wealth. But the culture and history of a country can impede the long-term legitimacy of the private property order and thus reverse the policy mix in a country. This leads us back to the project of the Friedmans—to educate not just the intellectuals of any society but also the citizenry. The economic liberal's project is not only to pursue a correct scientific understanding of the world but to change the intellectual climate of opinion toward an appreciation of the liberal project of constraining the government and encouraging the voluntary cooperation of individuals in society.[16] And here we can hope that the peoples of the former socialist economies will continue to benefit from the teachings of the Friedmans. In 1980, they ended *Free to Choose* with an optimistic chapter pointing out that "The Tide Is Turning." The call to action that the Friedmans make is characteristically straightforward:

> Needless to say, those of us who want to halt and reverse the recent trend should oppose additional specific measures to expand further the power and scope of government, urge repeal and reform of existing measures, and try to elect legislators and executives who share that view. But that is not an effective way to reverse the growth of government. It is doomed to failure. Each of us would defend our own special privileges and try to limit government at someone else's expense. We would be fighting a many-headed hydra that would grow new heads faster than we could cut the old ones off.
>
> Our founding fathers have shown us a more promising way to proceed: by package deals, as it were. We should adopt self-denying ordinances that limit the objectives we try to pursue through political channels. We should not consider each case on its merits, but lay down broad rules limiting what government may do. (Friedman and Friedman 1980, 287)

The danger is when a people forget that one of the most basic truths in human affairs is that the greatest threat to our freedom and our ability to realize peaceful social cooperation is the concentration of power in the hands of few. As the Friedmans warned, we had deluded ourselves in the twentieth century into believing that the concentration of power was not a threat as long as that power was to be used for good purposes (Friedman and Friedman 1980, 297). The costs to humanity were great, and nowhere as great as to those peoples who had to endure the good intentions of communism.[17] Hopefully, the reformers-turned-political-leaders learned not only how to privatize their economies but also how to constitutionally constrain their governments from the wisdom of Milton Friedman.

CONCLUSION

We live in a world where activists take to the streets to protest globalization and the inhumanity of capitalism, and at the same time are wearing sneakers constructed in Indonesia, sweaters made in England, pants made in the United States, and gas masks made in Canada. Of course, a free market economist himself can find fault with the International Monetary Fund, World Bank, and World Trade Organization, but that degree of subtlety is absent from our protestors' argument. On the other hand, they contend that the spread of markets throughout the world generates a race to the bottom in terms of labor policy and environmental control, and reinforces an unequal distribution of resources between rich and poor nations.

The logic of economic reasoning and the evidence point in the opposite direction. Markets are the most effective means available to improve the lot of mankind by spreading the international division of labor and increasing the productive capacity of mankind. Increases in real income can result only from increases in real productivity, and increases in real productivity result from improvements in labor skill, advancements in the stock of technological knowledge, and more effective management and organization of economic production within enterprises. Globalization brings all three of these sources of real productivity gains from the more developed economies to the less developed ones. Moreover, the expansion of the market area erodes the power of local monopolies and exposes political leaders to world standards of acceptable public policy toward the least advantaged in a society. As the Friedmans contended in *Free to Choose*, letting the market genie out of the bottle destabilizes the monopoly on political power that the ruling elite possess in a command economy.

"A tide of opinion, once it flows strongly, tends to sweep over all obstacles, all contrary views" (Friedman and Friedman 1980, 272). The events of the world subsequent to the publication of *Free to Choose* demonstrate the veracity of this claim. Communism collapsed, development planning in the Third World was rejected, and even the welfare state in Western democracies has resulted in fiscal reforms. But there are still those who agitate for more government involvement in the economy in the name of security—personal, economic, and national. For those of us who are persuaded by the argument that a people that is willing to trade off its liberty for security deserves neither, the demand for, and popularity of, these security measures is a disturbing trend. But perhaps we might find hope from a region of the world that in fact used to represent one of the main causes of our security concern in the West: the former communist economies. As Mats Lars, former prime minister of Estonia, remarked recently in describing the intellectual climate of opinion in Europe, "The most left wing parties in the new Europe, from an economic policy standpoint, are more right wing than the most right wing party from the Old Europe."[18] As I have tried to suggest here, the prevailing climate of public policy in East and Central Europe and the former Soviet Union that pushes for market initiative instead of government-provided security is in no small measure due to the powerful message conveyed in the works of Milton and Rose Friedman and their commitment to sound economic reasoning and to the philosophy of limited government.

NOTES

[1] This is actually the hypothesis that is most plausible in the face of the evidence, as I have argued in Boettke (1993, 3–4, and 2001, 1–6). Also see Boettke, ed. (2000) for an examination of the century-long debate among economists on the theory and practice of socialism.

[2] I take particular delight in this regard from the cartoon from the *Christian Science Monitor*, and reproduced in the Friedmans' memoirs, of a statue of Lenin being replaced with a statue of Milton Friedman in Poland. See Friedman and Friedman (1998, 513).

[3] Milton Friedman has stated on many occasions that he actually considers *Capitalism and Freedom* the superior book to *Free to Choose*, so the puzzle of the different receptions cannot be explained by reference to quality.

[4] The translations that I could find were as follows: French, 1980; German, 1980; Japanese, 1980; Norwegian, 1980; Spanish, 1980; Swedish, 1980; Danish, 1981; Italian, 1981; Portuguese, 1981; Chinese, 1982; Finnish, 1982; Hebrew, 1988; Polish, 1988, 1994; Czech, 1992; Estonian, 1992.

[5] The focus on the informational role of the price system was an aspect of Hayek's economic thought that was widely accepted from the 1940s. However, the development of "information economics" would not occur until the 1960s, after George Stigler's seminal paper on the topic. Moreover, while Friedman and Hayek were close intellectual allies in the battle against socialism, especially in their work with the Mont Pelerin Society, the emphasis on spontaneous order is more identified with Hayek than Friedman. But see the discussion in *Free to Choose* where they discuss language, science, and culture as examples of how complex and sophisticated orders can arise as the unintended consequence of individuals pursuing their own interests. See Friedman and Friedman (1980, 16–19).

[6] A comparison of the scientific impact of Hayek and Friedman, however, weights strongly in favor of Friedman. When I did *The Intellectual Legacy of F. A. Hayek* volumes for the Hayek centennial celebration in 1999, a citation study was conducted, and Friedman dominates over all the classical liberal economists who have won the Nobel Prize (Buchanan, Coase, and Stigler) and the older generation of Mises and Knight. See Boettke (1999, xi–xvi).

[7] It is important to stress two parts about the Friedmans' analysis. First, they never fell into the trap of believing the Soviet system was a textbook model of central planning that could allocate resources efficiently. Textbook central planning was impossible, but what emerged was a government-regulated market economy that served particular vested interests of the ruling elite (see Friedman and Friedman 1980, 1–2). Second, while recognizing the military prowess and threat of communism, the Friedmans never bought into the economic, let alone moral, superiority of the communist system like so many of their contemporaries. Economists such as John Kenneth Galbraith and Paul Samuelson wrote well into the 1980s about the productive capacity of the socialist system to outperform the capitalist system. In this respect, the analysis by the Friedmans of the rotting of the socialist system from within and the instability that would be introduced once minor market reforms were implemented was prescient indeed. On the development of the implications of the noncentrally planned nature of the Soviet system, see Roberts (1971) and Boettke (1990, 1993, and 2001). On how the introduction of market reforms and inconsistent policy shifts caused the downfall of the communist regime under Gorbachev, see Boettke (1993).

Milton and Rose Friedman's "Free to Choose"

8 We learn in their memoirs that the Friedmans had a fascination with learning up close the operation of different economic systems, including their year-long trip to visit over twenty countries in the early 1960s (Friedman and Friedman 1998, 279–332). The Friedmans also were involved in the effort to bring the latest teachings of market economics to economists in the former Soviet bloc, dating from the mid-1960s. These meetings were held under the auspices of an Italian research center (CESES) under the direction of Renato Mieli, who worked in cooperation with G. Warren Nutter (Friedman and Friedman 1998, 338). Nutter was Milton Friedman's student at the University of Chicago and challenged the dominant opinion among Sovietologists at the time that despite whatever political problems the Soviet system confronted, the economy had performed admirably in lifting a country from a backward peasant economy to an industrialized economy in less than a generation. Nutter's (1962) work challenged the growth accounting that was being done on Soviet-type economies, and his revised figures called for a reassessment of the economic prowess of the Soviet system. In other work, Nutter (1983) also challenged the notion that either shadow pricing mechanisms or the decentralization of administration could provide the incentives and information required to improve the economic performance of the Soviet-type economy. Markets without private property, Nutter argued forcefully, were a grand illusion, and without the establishment of private property rights, Soviet reforms were bound to lead to frustrating results. Nutter was a trailblazer whose works were often dismissed during his day, only to have them viewed as singularly prescient after the collapse of the Soviet Union in 1991. For a discussion of the debate over Soviet economic growth figures, see Boettke (1993, 12–45).

9 Gaidar, in an interview for the PBS series "Commanding Heights," states in response to a question about Milton Friedman's influence: "Yes, I read Friedman's books with interest, and also Hayek. They were very authoritative for us, but all the same far away from our domestic realities."

10 Of course, this is the criticism of the insider deals that were struck in Russia that have led to a generation of oligarchs. See Goldman (2003). Contrary to these arguments, I would stress with Boyko, Shleifer, and Vishny (1995) that the primary purpose of privatization was "destatization," and reform efforts should be judged against that standard. The lingering economic problems in East and Central Europe and the former Soviet Union are a consequence not of either "insider" or "voucher" privatization but the lack of actual implementation of reforms, the partial nature of many of the reforms that were implemented, and the continuing interference of the state in economic life.

11 Friedman (1990, 91) points to changes in communications, such as e-mail, telephone, and fax machines, as effectively eroding the monopoly power of the U.S. Post Office. On how this example of the U.S. Post Office can serve as a general model for transition economies in privatization, see Boettke and Leeson (2003).

12 See Burba (1999) for a discussion of Balcerowicz's advocacy of the flat tax and the origination of the idea in Friedman's *Capitalism and Freedom* (1982, 172–76).

13 The empirical record of postcommunism is often clouded by (1) an unreliable base state, as the official data on the economy at the time of the collapse of communism often overestimates the economic health of these economies, (2) failure to adequately account for the problems of a shortage economy, repressed inflation, negative value-added production techniques, etc., that characterized the Soviet-type economy and thus misinterprets the initial period of the transition

in terms of price adjustments and the reallocation of labor and capital, and (3) failure to incorporate into the analysis the unofficial economy that emerges as individuals attempt to evade the regulations, registration fees, and taxation of the official system even after so-called reforms have been implemented. These problems are discussed in Boettke and Leeson (2003).

[14] Personal interview by Peter Boettke, Scott Beaulier, and Susan Anderson with Vladimir Dlouhy at his office in Prague on July 14, 2003.

[15] The results obtained from both the Index of Economic Freedom and the Economic Freedom Index are clear on this: Economic freedom (security of private property, freedom of pricing, sound money, fiscal responsibility, low taxes, nonintrusive regulation, and open international trade) is positively correlated with economic growth. It is just not the case that countries can score low on these different indices of economic freedom and experience significant economic growth and improvements in the standard of living of the average citizens in their country. An online description of the Economic Freedom Index is available at www.freetheworld.com.

[16] As stressed by Milton Friedman, the classical liberal economist will do a disservice to his cause if he allows his ideological position to crowd out his positive economic analysis. Instead, the case for classical liberalism must be grounded in sound economic analysis. To accomplish this, the economist should in the first instance engage in a positive analysis of any policy proposal by treating the ends of that proposal as public-spirited and unobjectionable from a broadly accepted moral standard (for example, we want to help the least advantaged in society). Then subject that proposed policy to economic analysis, with the surprising result that much of what is advocated in the name of the public interest actually fails to promote that objective. Confronted with this knowledge of the conclusions of positive economic analysis, politicians should abandon their previous policy and pursue one more suited to meet the stated objectives. In the face of repeated failures to respond to the logic and evidence provided by sound economic analysis, Friedman then suggests that the economist can offer a positive analysis of the political process of policymaking. At this level, the logic and evidence lead one to expose how special interest groups operate in concert with politicians and a permanent bureaucracy to concentrate benefits on the well-organized and disperse benefits on the unorganized and establish effective constraints on any attempt to change the status quo. Again, values are not unwarrantedly imported into the analysis. The conclusions that emerge about the inertia of the status quo and the logic of concentrated benefits and dispersed costs are derived through sound economic reasoning. Finally, in building the positive construction of a constitution of economic policy, Friedman argues that we need to learn from thinkers such as Adam Smith and Thomas Jefferson and build institutional constraints that will enable "bad men" to do least harm, rather than free "good men" to accomplish all that they intend. *Free to Choose* is a perfect illustration of how to engage in each of these levels of analysis in political economy.

[17] See R. J. Rummel (1994) and Courtois et al. (1999).

[18] Remarks at the graduation ceremonies of the American Institute for Political and Economic Studies (Fund for American Studies) at Charles University, August 2, 2003.

REFERENCES

Aganbegyan, Abel. 1989. *Inside Perestroika: The Future of the Soviet Economy*. New York: Harper and Row.

Aslund, Anders. 2002. *Building Capitalism: The Transformation of the Former Soviet Bloc.* New York: Cambridge University Press.

Boettke, P. 1990. *The Political Economy of Soviet Socialism: The Formative Years, 1918–1928.* Boston: Kluwer Academic Publishers.

———. 1993. *Why Perestroika Failed: The Politics and Economics of Socialist Transformation.* New York: Routledge.

———. 1999. "Which Enlightenment, Whose Liberalism? F. A. Hayek's Research Program for Understanding the Liberal Society." In Peter J. Boettke, ed., *The Legacy of F. A. Hayek: Politics, Philosophy, Economics*, 3 volumes. Cheltenham, UK: Edward Elgar, Vol. 1, xi–lv.

———. 2001. *Calculation and Coordination: Essays on Socialism and Transitional Political Economy.* New York: Routledge.

Boettke, P., and Peter Leeson. 2003. "Is the Transition to the Market Too Important to Be Left to the Market?" *Economic Affairs* 23 (1), March.

Boettke, P., ed. 2000. *Socialism and the Market Economy: The Socialist Calculation Debate Revisited*, 9 volumes. New York: Routledge.

Boyko, M., Andrei Shleifer, and Robert Vishny. 1995. *Privatizing Russia.* Cambridge, MA: MIT Press.

Buchanan, James. 1997. *Post-Socialist Political Economy.* Cheltenham, UK: Edward Elgar.

Buchanan, James, and Gordon Tullock. 1962. *The Calculus of Consent.* Ann Arbor, MI: University of Michigan Press.

Burba, Andrzej J. 1999. "Emerging Leader of the Tax Avant-Garde: Poland's Proposal to Institute a Flat Tax on Individual and Corporate Incomes." *Vanderbilt Journal of Transnational Law* 32 (5).

Coase, Ronald. 1992. "The Institutional Structure of Production." *American Economic Review* 82 (September): 713–19.

Courtois, S., et al. 1999. *The Black Book of Communism: Crime, Terror, Repression.* Cambridge, MA: Harvard University Press.

Freeland, C. 2000. *Sale of the Century: Russia's Wild Ride from Communism to Capitalism.* New York: Crown Publishers.

Friedman, M. 1982. *Capitalism and Freedom.* Chicago: University of Chicago Press, 1962.

———. 1984. *Market or Plan?* London: Centre for Research into Communist Economies.

———. 1990. *Friedman in China.* Hong Kong: Chinese University in Hong Kong Press.

Friedman, M., and Rose D. Friedman. 1980. *Free to Choose: A Personal Statement.* New York: Harcourt Brace Jovanovich.

———. 1998. *Two Lucky People: Memoirs.* Chicago: University of Chicago Press.

Goldman, Marshall I. 2003. *The Piratization of Russia: Russian Reform Goes Awry.* New York: Routledge.

Hayek, F. A. 1944. *The Road to Serfdom.* Chicago: University of Chicago Press.

———. 1960. *The Constitution of Liberty.* Chicago: University of Chicago Press.

———. 1973. *Law, Legislation and Liberty: Vol. 1.* Chicago: University of Chicago Press.

———. 1976. *Law, Legislation and Liberty: Vol. 2.* Chicago: University of Chicago Press.

———. 1979. *Law, Legislation and Liberty: Vol. 3.* Chicago: University of Chicago Press.

North, Douglass. 1994. "Economic Performance Through Time." *American Economic Review* 84 (June): 359–68.

Nutter, G. Warren. 1962. *The Growth of Industrial Production in the Soviet Union.* Princeton, NJ: Princeton University Press.

———. 1983. *The Political Economy of Freedom.* Indianapolis, IN: Liberty Fund.

Pejovich, S. 2003. "Understanding the Transaction Costs of Transition: It's the Culture, Stupid." *Review of Austrian Economics* 16 (4): 347–61.

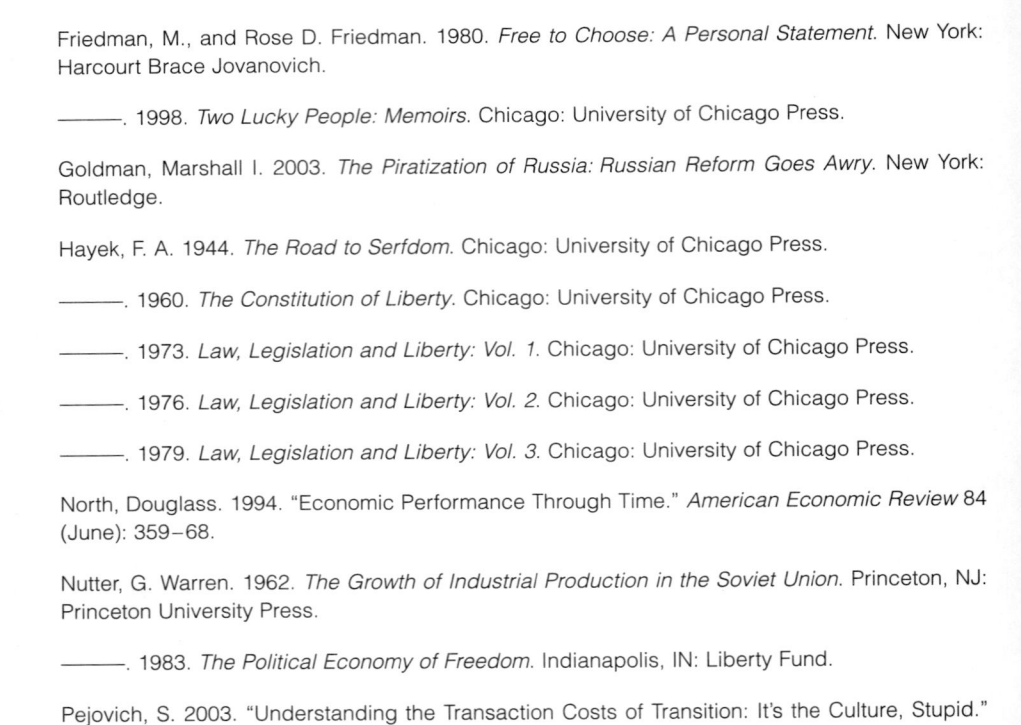

Roberts, P. C. 1971. *Alienation and the Soviet Economy.* Albuquerque, NM: University of New Mexico Press.

Rummel, R. J. 1994. *Death by Government.* New Brunswick, NJ: Transaction Press.

Sachs, Jeffrey. 1993. *Poland's Jump to the Market Economy.* Cambridge, MA: MIT Press.

Free to Choose in China

Gregory C. Chow

What have been the significant changes in China's economic and political institutions? I will answer this question by discussing eight topics chosen from Milton and Rose Friedman's book *Free to Choose*. The topics are economic freedom, the relation between economic and political freedom, the role of government, social welfare, education, consumer protection, macroeconomic policy, and trends in the development of economic and political freedom.

ECONOMIC FREEDOM IN THE LAST HALF CENTURY

There have been significant changes in economic freedom in China in the last half century. Economic freedom began to be severely restricted when central planning was introduced in 1953. It has increased steadily since economic reform started in 1978. Today there is much economic freedom in China.

The Chinese government adopted Soviet-style central economic planning in 1953 when it initiated the First Five-Year Plan of 1953–57. For a quarter of a century that followed, freedom in production, distribution, and consumption was restricted. For industrial production, private enterprises were converted into state-owned enterprises, which had to fulfill output targets approved by central planning. Materials used in production were centrally allocated. Urban workers were assigned jobs in the state enterprises and could not move from city to city.

In agriculture, the commune system was established in 1958. Private farming was abolished. The farmers in a traditional village were organized as a team in a commune to farm collectively. Free trade of farm products was abolished. Rural markets were banned. Each team in a commune was assigned quotas of output to be delivered to a government procurement agency for distribution in urban areas. A system of rationing of consumer goods in urban areas was put in place. Each consumer was given coupons to buy specified amounts of food grain, oil, eggs, and cloth. Retail stores were operated by the government. There was no free market for housing. Housing units were assigned to employees by their work units at very low rents. Privately owned automobiles were nonexistent.

China's door was closed to the outside world. Foreign trade was handled by the government, which determined the exports and imports of specific products to and from specific countries. The main trading partners were the Soviet bloc countries. Foreign investment from Western countries was not welcome. Chinese citizens were not allowed to travel abroad. The border with Hong Kong, which had been open without any restrictions, was closed in the early 1950s.

Recognizing the shortcomings of central economic planning from years of experience and witnessing the economic success of neighboring market economies of Hong Kong, Singapore, Taiwan, and South Korea, the Chinese government, under the leadership of Deng Xiaoping, began economic reform toward a market-oriented economy in 1978. In agriculture, private farming was revived. The commune system was abolished by 1982. This change occurred through initiatives from below when some commune leaders discovered that output quotas could be met by assigning a piece of land to each farm household to farm and collecting a quota of output from it, rather than by having all farm households in a team to farm collectively. Under this "household responsibility system" output increased significantly because the farmers could reap the fruits of their additional effort. The central government soon adopted this system as national policy. Increase in farm output allowed the gradual abandonment of the rationing of consumer goods.

Urban industrial reform took several steps that were introduced in a period of about two decades, from granting autonomy to state enterprises in production decisions to converting them to shareholding corporations traded in stock markets in the late 1990s. China's door was opened to foreign trade and investment. In 2001, China joined the World Trade Organization to open its door further by lowering tariffs and allowing more foreign competition in agriculture, manufacturing, and service industries and in domestic trade. Observers attribute the success of China's economic reform to the gradual and experimental approach taken by the leaders, who were pragmatic and adopted what worked without being subject to ideological constraints. As Deng advised, "Seek truth from facts."

Today China has a well-functioning market economy in spite of its shortcomings. Economic freedom of the citizens has been greatly enhanced. Private enterprises have flourished. Rationing has been abolished since the early 1980s, and consumer goods are abundant. Housing has been privatized. The purchase of new automobiles in 2003 is to exceed 4.2 million units. The Chinese people can travel freely both inside and outside the country, many having come to the United States to study. They can choose and change their jobs fairly freely, although many do not move because of the benefits of entitlements under the welfare system administered by state-owned enterprises.

There are unions, but there is no union power that restricts the freedom of employers to choose workers and the freedom of nonunion workers to choose their jobs. (In this respect, China has more economic freedom than the United States, and the topic of union power is not included in the eight topics of this paper.) There appears to be no serious infringement of economic freedom in China, with the exception of the one-child policy that allows only one child for an urban family and an additional child for a rural family if the first child is a girl.

RELATION OF ECONOMIC FREEDOM AND POLITICAL FREEDOM

On pages 2–3 of *Free to Choose* we read: "Economic freedom is an essential requisite for political freedom. By enabling people to cooperate with one another without coercion or central direction, it reduces the area over which political power is exercised. In addition, by dispersing power, the free market provides an offset to whatever concentration of political power may arise." China provides many examples for these observations.

As private farming under the household responsibility system replaced collective farming under the commune system, agricultural output increased in the early 1980s and labor was free to move. In a trip along the Yangtze River to see the Three Gorges in 1982, I witnessed many farmers on the boat carrying farm products to neighboring areas for sale and farmworkers carrying carpenter tools to find work elsewhere. As economic freedom increased, the administration of the commune system ceased to function.

There remained the need to provide security, to protect public land, and to attend to the public affairs of rural villages that were formerly within the domain of the commune system. This need, together with the economic interest of and economic power acquired by the farmers, provided the primary reasons for the direct elections of village officials that have become widespread in rural China. The central government endorses such elections because the elected officials perform important functions in Chinese villages. Village elections in China are a major component of the change of China's political institutions toward a more democratic government.

As more consumer goods became available, rationing was no longer necessary. Goods began to be distributed in rural markets and in collectively or privately owned stores in urban areas. The role of government procurement and trading is greatly reduced. Services formerly provided by employees in government-owned retail stores, hotels, train stations, theaters, and other service-providing establishments are now provided by private enterprises for profit. The quality of services has greatly improved. The sphere of government influence has been reduced in all aspects of economic life, including production, distribution, employment, foreign trade, and foreign investment. Government bureaucrats are replaced by nongovernment employees who are more service-minded. All this, as the Friedmans say, "reduces the area over which political power is exercised."

The widespread abuse of economic power under the previous system of government ownership of assets and government control of resource allocation has been reduced with the rise of the free market. The system of economic planning itself induced the Chinese to take full advantage of the assets under their control. Under central planning, major economic assets were owned and controlled by the government, but in reality the government had to assign people to control and use the assets on its behalf and in the name of the state. The people who had the power to manage government assets used them for their own benefit. Corruption was only one example when the bureaucrats controlling some economic assets extracted money from people who desired to use them. A driver of a government-owned car could use the car for personal benefit. If another person desired to use the car, he would have to compensate or appease the driver since no taxicabs were available.

Under this system, the Chinese people became frustrated when they had to beg to get served or to acquire the essential consumer goods. They then aired their frustrations and returned the favor to others when other people desired goods and services from them. The quality of services provided in China was poor in general. People were unkind to one another whenever one person needed something from someone who had control of the needed asset or service. Barters became widespread. A person in charge of selling low-price and scarce theater tickets could exchange the tickets for scarce consumer goods distributed in government stores.

With the appearance of the market economy, the quality of services provided by the Chinese people has gradually improved, and the people have been kinder to one another. Now money can be used to purchase goods and services. Fewer people connected with the government have monopoly power over the control of economic resources that others need.

As collective and private enterprises flourished, a group of well-to-do citizens has emerged and gained influence and economic power in the Chinese society. In the late 1990s, under the leadership of General Secretary Jiang Zemin, the Chinese Communist Party began to accept capitalists into its membership. China has a one-party system. There are other political parties, but none can control the Chinese government. They exercise their political influence mainly through a National Political Consultative Conference, which represents diverse political interests and meets regularly at the same time as the National People's Congress. The recommendations of the conference are seriously considered by the People's Congress. Political power of the people is exercised by indirect election of members of the People's Congress and of the members of the Communist Party Central Committee (the latter by members of the party only). To the extent that membership of the Communist Party is more open, more people will have more political freedom and more opportunities to participate in government affairs.

As economic freedom has increased, so has political freedom, although it is difficult to ascertain the precise effect of the former on the latter. Political freedom is in principle guaranteed in the Chinese Constitution adopted by the People's Congress on December 4, 1982. In Chapter I, "General Principles," Article 2 states: "All power in the PRC belongs to the people. The organs through which the people exercise state power are the National People's Congress and the local people's congresses at different levels." Article 28 states: "The state maintains public order and suppresses treasonable and other counter-revolutionary activities; it penalizes actions that endanger public security and disrupt the socialist economy ...." In Chapter II, "The Fundamental Rights and Duties of Citizens," Article 35 provides all citizens with "freedom of speech, the press, of assembly, of association, of procession and demonstration," while Article 36 provides "freedom of religious belief." However, the stated freedoms can be restricted by appealing to Article 28 of the Constitution.

In reality, the Chinese people do have much more freedom than before. They can talk freely in private gatherings and even openly in professional meetings without fear of being prosecuted. For instance, a Chinese economics professor openly criticized the labor theory of value in Marxian economics in a paper presented before a conference in Beijing in 1999. There is freedom of the press to a considerable extent, as the nongovernment press has expanded rapidly in recent years and attracted a large readership. This includes daily or weekly newspapers, magazines, and books. Opinions expressed therein are open and free, subject to only a minor degree of censorship. Censorship of foreign books is almost nonexistent.

Information available to the public is somewhat restricted because the government has control over TV and radio stations as well as the Internet. However, the control is limited because the Chinese have access to shortwave radios, and it is difficult to control the use of fax machines and the flow of information through the Internet. People residing near Hong Kong can get access to TV stations in Hong Kong that are mainly private and free.

Religious freedom has increased, as illustrated by the rapid increase in attendance in Christian churches and worship in Buddhist temples. Church attendance has been influenced by the efforts of overseas Chinese Christians, especially those in Hong Kong who invest in China's economic and educational institutions and sometimes also engage in religious activities.

Government control of the press has been reduced partly because the nongovernment press has spread by the increase in demand for reading material in an affluent society. Conversely, demand for government-printed material has declined. As the August 5, 2003, issue of *People's Daily* reports, "Party and government-run newspapers and magazines will be forced to commercialize or face closure under major reforms.... State administration statistics show that there were 2,137 newspapers in China last year, but newspapers relying on administrative orders for subscribers accounted for 40 per cent of the total." Government and party organizations were asked to close their newspapers or magazines if they did not have sufficient paying subscribers.

Under a one-party political system, which limits political freedom, there is much economic freedom in China, for political freedom is not a necessary condition for economic freedom. Economic freedom will help increase political freedom, but political freedom may not help increase, and can sometimes reduce, economic freedom, as illustrated by many examples in the United States that have been documented in *Free to Choose*. There is much political freedom in the United States, but the many infringements of economic freedom are the subject of *Free to Choose*.

THE ROLE OF GOVERNMENT

On page 5 of *Free to Choose*, we read: "The view that government's role is to serve as an umpire to prevent individuals from coercing one another was replaced by the view that government's role is to serve as a parent charged with the duty of coercing some to aid others."

China's experience illustrates very well "the government's [proper] role as an umpire to prevent individuals from coercing one another." In China, the need for the government to provide law and order is very important because law and order is important for social stability and economic progress.

As someone accustomed to law and order in the United States since 1948, it has taken me several visits to China to appreciate its importance. Several experiences impressed me. In 1982, when I tried to send a telegram in a post office in Guangzhou, I found that people did not line up in front of the service window and there was no way for me to get to the front to submit the draft of my telegram. I had to let a Chinese friend fight his way through the crowd to send the telegram. I wished an officer had been there to guide people to line up.

As a second experience, my wife and I were provided a tour guide while visiting Confucius' Temple in Shandong province in 1985. As the guide was explaining the points of interest to us, people began to crowd in and surrounded the guide, to the point that Paula and I were too far separated from him to hear what he was saying. As a third example, I was traveling by car to visit the site of Yuan Ming Yuan in Beijing. As we approached the site, we found a roadblock set up illegally by local residents to collect tolls. Our driver had to pay the toll before he could drive through. Extraction of fees of all kinds by local residents from travelers passing through their territories was and is quite common in China. A strong government is needed to prevent some citizens from extracting rents illegally from others.

The Chinese government has been urged to protect intellectual property rights, which are violated in China in the sale of pirated CDs of computer software, music, and movies. Many Chinese regard law and order as more important than freedom and desire a strong government to protect them from coercion by others.

The second and improper role of the government—"to serve as a parent charged with the duty of coercing some to aid others"—can be illustrated by many examples in the early years of the People's Republic of China (PRC). Private land was redistributed in the early 1950s to the farmers. Private enterprises were turned to state-owned enterprises. Agricultural output produced in the communes was taxed for distribution to the urban population. Such coercions have been reduced with the introduction of market institutions where private property is respected. At the same time, the growing importance of the first, legitimate role of the government—"as an umpire to prevent individuals from coercing one another"—signifies that the political system in China has improved.

SOCIAL WELFARE

During the period of central planning, job security was guaranteed, and support for the aged who were not supported by their children was provided by the communes in rural areas and by the state-owned enterprises and other government institutions to their employees in the urban areas. Health care in the entire country was provided under an efficient three-tier system, with village doctors treating simple illnesses in village stations, physicians with three years of medical training after high school in health centers, and better trained doctors in city hospitals in turn taking care of more difficult illnesses. A community-financed Cooperative Medical System (CMS) funded and organized health care for almost the entire rural population. Health centers and hospitals associated with state-owned enterprises and other government institutions cared for employees and their family members.

With the introduction of economic reform, the commune system broke down, and state-owned enterprises were made financially independent and downsized. Private farmers had to find their own work and support themselves in old age. Urban workers could become unemployed. In rural areas, agricultural reforms in the early 1980s led to the disintegration of the cooperative organizations that formed the basis of CMS. Rural populations had to finance their own health care, while many village doctors had their own private practice. In the urban areas, state enterprises and other government organizations had difficulty financing the health care of their employees.

Since the mid-1990s, the Chinese government has attempted to set up, step by step, a nationally unified social security system for the urban population, under the central management of the labor and social security administration departments and with social insurance funds partly contributed by the central government. Labor and social security departments at all levels are responsible for the collection, management, and payment of the social insurance funds. Besides contributions from employers and employees as stated below, the central government allocated 98.2 billion yuan in 2001 for social security payments, 5.18 times the amount in 1998 as it was expanding the system to cover larger segments of the population in steps. (All statistics on the development of the social security system can be found on the web site of *People's Daily*, http://english.peopledaily.com.cn/, under "White Paper on Labor and Social Security in China" in the section "White Papers of Chinese Government.")

In 1997, a uniform, basic old-age insurance system for enterprise employees was established, financed by 20 percent of the enterprise wage bill and 8 percent of the employee's wage. A part of the premiums from enterprises goes to mutual assistance funds and the rest to personal accounts, while the premiums from the employees go entirely to personal accounts that belong to the employees themselves and can be inherited. Employees participating in this program increased from 86.71 million in late 1997 to 108.02 million at the end of 2001, while the number receiving pensions increased from 25.33 million to 33.81 million, with the average monthly basic pension per person increasing from 430 yuan to 556 yuan. The rural population pay their own insurance premiums and withdraw funds from personal accounts with subsidies from the government.

In 1999, an unemployment insurance system was introduced, financed by 2 percent of the wage bill paid by employers and 1 percent paid by employees. Unemployment insurance benefits are lower than the minimum wage but higher than the minimum living allowance guaranteed for all laid-off workers. The period of drawing insurance depends on the length of the period in which insurance payments have been paid, with twenty-four months as the maximum. The number of insured persons increased from 79.28 million in 1998 to 103.55 million in 2001.

On health care, important policies were announced on January 15, 1997, in the "Decision on Health Reform and Development by the Central Party Committee and State Council." The basic objective of the Decision is to ensure that every Chinese will have access to basic health protection. For the rural population, the strategy is to develop and improve CMS through education, by mobilizing more farmers to participate, and gradually expanding its coverage. For urban employees, a basic medical insurance system was established in 1998, financed by 6 percent of the wage bill of employing units and 2 percent of the personal wages. By the end of 2001, 76.29 million employees had participated in basic insurance programs. In addition, free medical services and other forms of health care systems covered over 100 million urban population.

In terms of saving for old age, rural populations in China have more freedom than urban populations to make their decisions but are subject to more risks. Urban populations have their own personal accounts, with amounts depending on their own contributions. Their old-age insurance system has features similar to the pension system in American universities, with both the employer and the employee contributing to the fund and with each employee having his personal account. Rural populations in China have more freedom to choose their work but are not guaranteed unemployment benefits, as are the urban workers.

EDUCATION

In China the government has controlled the education system since the early 1950s, but in recent years the role of nongovernment-operated and financed education at all levels has become very important.

When the government decided to adopt Soviet-type economic planning in the early 1950s, the system of higher education was modeled after the Soviet Union's, along with economic planning. The government seized control of all educational institutions. Private universities were closed and converted into public educational institutions. Liberal education ceased to exist. Education served mainly the purpose of economic development. For this purpose, it was believed that a university student should concentrate on one subject, rather than receiving a general liberal education, and that each university should specialize also. Former universities, public and private, were reorganized. One university was divided into several, more specialized institutions. The school of arts and sciences, the medical school, the engineering school, and the school of agriculture of one university were separated and became colleges on their own. Each government ministry responsible for the production and distribution of one major product had under its control colleges to train people to work in a specialized area. This was like having the School of Mines under the Bureau of Mines in the United States repeated many times for different industrial ministries. The People's Bank administered a graduate school to train staff for the Bank and its branches in different provinces. The People's (Renmin) University was established to train government officials.

At lower educational levels, the government also controlled all schools. Being welfare-minded, it wanted to raise the level of education for the entire population. By 1978, the literacy rate among the population 15 years of age or over was raised to 82 percent. No private schools were allowed. To see the expansion in education, refer to Table 1 on school enrollment at different educational levels. The enrollment figures have not been adjusted for the increase in population in the corresponding school age.

After 1978, Deng Xiaoping initiated economic reform toward a market-

Table 1
Student Enrollment by Level of School, 1949 to 1981[a]
(10,000 Persons)

			Secondary Schools[b]		
Year	Total	Institutions of Higher Learning	Secondary Specialized Schools	Regular Secondary Schools	Primary Schools
1949	2,577.6	11.7	22.9	103.9	2,439.1
1950	3,062.7	13.7	25.7	130.5	2,892.4
1951	4,527.1	15.3	38.3	156.8	4,315.4
1952	5,443.6	19.1	63.6	249.0	5,110.0
1953	5,550.5	21.2	66.8	293.3	5,166.4
1954	5,571.7	25.3	60.8	358.7	5,121.8
1955	5,788.7	28.8	53.7	390.0	5,312.6
1956	6,987.8	40.3	81.2	516.5	6,346.6
1957	7,180.5	44.1	77.8	628.1	6,428.3
1958	9,906.1	66.0	147.0	852.0	8,640.3
1959	10,489.4	81.2	149.5	917.8	9,117.9
1960	10,962.6	96.2	221.6	1,026.0	9,379.1
1961	8,707.7	94.7	120.3	851.8	7,578.6
1962	7,840.4	83.0	53.5	752.8	6,923.9
1963	8,070.1	75.0	45.2	761.6	7,157.5
1964	10,382.5	68.5	53.1	854.1	9,294.5
1965	13,120.1	67.4	54.7	933.8	11,620.9
1966	11,691.9	53.4	47.0	1,249.8	10,341.7
1967	11,539.7	40.9	30.8	1,223.7	10,244.3
1968	11,467.3	25.9	12.8	1,392.3	10,036.3
1969	12,103.0	10.9	3.8	2,021.5	10,066.8
1970	13,181.1	4.8	6.4	2,641.9	10,528.0
1971	14,368.9	8.3	21.8	3,127.6	11,211.2
1972	16,185.3	19.4	34.2	3,582.5	12,549.2
1973	17,096.5	31.4	48.2	3,446.5	13,570.4
1974	18,238.1	43.0	63.4	3,650.3	14,481.4
1975	19,681.0	50.1	70.7	4,466.1	15,094.1
1976	20,967.5	56.5	69.0	5,836.5	15,005.5
1977	21,528.9	62.5	68.9	6,779.9	14,617.6
1978	21,346.8	85.6	88.9	6,548.3	14,624.0
1979	20,789.8	102.0	119.9	5,905.0	14,662.9
1980	20,419.2	114.4	124.3	5,508.1	14,627.0
1981	19,475.3	127.9	106.9	4,859.6	14,332.8

[a] Excludes spare-time schools.
[b] Excludes workers' training schools.
SOURCE: *Statistical Yearbook of China*, 1981, 451.

oriented economy. Education was an important part of this reform process. The system of higher education was gradually changed. The main direction was to abandon the Soviet-style higher education system introduced in the 1950s in favor of a more comprehensive and integrated system, as practiced in the 1940s. For education at all levels, the government has allowed "citizen-operated" schools to develop and flourish side by side with the schools administered by government at all levels.

The gradual change occurring in education reform, as in economic reform, has taken two and a half decades and is still incomplete, but in both cases we can see what has been accomplished. Higher education has become less specialized, with universities reorganized by including previously separated colleges of medicine, engineering, and agriculture and other functional disciplines. While the Ministry of Education in Beijing still controls thirty-some major universities, the remaining state universities are under the control of the governments of provinces, cities, and townships. The curricula, including economics in particular, have changed to suit the working of a market economy.

Educational institutions at all levels continued to improve not only through the efforts of the central, provincial, and local governments but also by the efforts of the nongovernment sectors. "Citizen-operated" or privately financed schools at all levels have become widespread because there is large demand for them as the Chinese people have become richer and because the schools can be profitable.

In the late 1980s, I visited a primary school near Guangzhou that was established privately. The parents had to pay 100,000 RMB, worth about $30,000 at the time, at the beginning of the first year for a six-year primary-school education for one child. The investors of this school used the income to build a building on a piece of land leased from the town government at low rent to encourage education. The school was said to be profitable. It was very good in terms of the quality of the teachers and the orderly behavior of the students. Often such schools were established formally by, or in the name of, an association. Associations of all forms sprang up rapidly in China after economic reform started. They are accorded some legal status that a private individual may not possess. They have already invested certain fixed costs in the right to use land or a building, in the establishment of some legal status, in the personal connections of its management and staff, and in the public recognition of the organization. All such investment can be exploited to sponsor a school or another kind of business enterprise.

Nongovernment schools have grown rapidly, not only because they are economically viable but also because many overseas Chinese are willing to support them. Chinese outside Mainland China have poured in money to support all kinds of education in China. Both financial resources and knowledge on administering educational institutions were supplied to China for its benefit, as in the case of foreign investment, except that investment in education is nonprofit in most cases. The investor contributes both time and money to improve education in China.

Observers have suggested that the Chinese education system is deficient partly because the government spends too little on education. They would cite statistics on the amount of government expenditure as a percent of GDP to support this claim. In 1995, public expenditure on education was only 2.5 percent of GDP in China, as compared with 5.4 percent in the United States and 5.2 percent as the world average. These statistics have not taken into account the nonpublic expenditures, contributions by overseas Chinese and other friends, and the spending by the parents to pay tuition in "citizen-operated" schools and in public schools. Outside contributions to education in certain towns, counties, and villages are substantial, including, in particular, some towns and villages near Hong Kong.

The importance of privately financed education in China and some other countries has been documented in a report, *Financing Education—Investments and Returns*, published in 2002 by the United Nations Educational, Scientific and Cultural Organization (UNESCO) and the Organization for Economic Cooperation and Development (OECD), which focuses on sixteen emerging economies. Funds from a wide range of private sources, including individuals and households, contribute much more to education in these countries than in the OECD member states. In Chile, China, and Paraguay, for example, more than 40 percent of the total amount spent on education comes from such private sources. The OECD average is 12 percent. There has been a rapid development of private education services in these countries, from wholly private, independent institutions to schools that have been subcontracted by governments to nongovernmental organizations. In China and Zimbabwe, government-subsidized, community-managed schools are said by the above report to be the backbone of the education system. (Author's Note: When I presented the 40 percent figure for private spending on education in China in the conference, Gary Becker questioned its accuracy. I then supplied some personal observations to support this figure. Even public schools in China at all levels collect tuitions that should be included in private spending, not to mention the large number of nonpublic educational institutions. Data on government and nongovernment funding of education are found in Table 20-35 of *China Statistical Yearbook 2003*.)

For the United States, the major concern expressed in *Free to Choose* for primary and secondary education is that the parents have to pay taxes to finance poorly operated public schools and do not have sufficient choice of schools. For college education there is too much government subsidy, while the students should pay their own expenses as a form of investment to prevent waste and misuse of educational resources. In China, if privately financed education is widespread and accounts for over 40 percent of total expenditures on education, the concern about public schools using up funds from parents who might prefer to send their children to private schools appears to be less pressing. Furthermore, in the case of primary and secondary school education the local governments provide choice of schools to the parents. There are usually several public schools of different qualities in the same area. Students can enter a good school if they can pass the required examinations. Furthermore, to the extent that the parents have to pay tuition to public schools that are not entirely financed by tax money, this is like the voucher system advocated by Milton Friedman, which provides parents the freedom to choose schools for their children.

For college education, in recent years government policy has been to increase tuition for students and to encourage universities to be more independent financially by seeking ways to generate income, including charging tuition for courses offered to professionals. This policy is consistent with putting the financial burden on students seeking higher education. Thus, the education system in China appears to address the concerns expressed in *Free to Choose* in the following two respects: Parents have considerable choice of private and public elementary and secondary schools for their children. Students receiving higher or professional education have to pay for it. At the same time, the government plays an active role in controlling and financing education at all levels.

CONSUMER PROTECTION

Chapter 7 of *Free to Choose*, on "Who Protects the Consumer?", includes the following statements: "The market must...be supplemented by other arrangements in order to protect the consumer from himself and from avaricious sellers, and to protect all of us from the spillover neighborhood effects of market transactions. These criticisms of the invisible hand are valid.... The question is whether the arrangements that have been recommended or adopted to meet them...are well devised for that purpose, or whether...the cure may not be worse than the disease" (189). "Every act of [government] intervention establishes positions of power. How that power will be used and for what purposes depends far more on the people who are in the best position to get control of that power and what their purposes are than on the aims and objectives of the initial sponsors of the intervention" (193).

In the process of reforming China's economy, the National People's Congress has passed much new legislation governing economic transactions. Of particular importance for consumer protection are the Product Quality Law of 1993 and the Law on the Protection of the Rights and Interests of Consumers of 1994. These laws can be found on the web site http://www.chinalaw114.com/englishlaw/category.asp?cate=31.

The Product Quality Law of the People's Republic of China stipulates that "producers and sellers are responsible for the product quality according to the provisions of the law" (Article 3). "It is forbidden to forge or infringe upon quality marks...the place of origin...factory names" (Article 4). The product quality supervision and control departments of the State Council are responsible for the supervision and control of the quality of the products" (Article 6). To ensure product quality, "enterprises may apply voluntarily for certification of their quality control systems to the product quality supervision and control departments" or "for certification of the quality of their products" (Article 9). "The State shall institute a system of supervision and chiefly random examination to check randomly samples of major industrial products which may be hazardous to health…" (Article 10). Chapter 3 spells out the responsibilities and obligations of producers and sellers to ensure the products are safe, have quality and other characteristics as specified, and are certified for quality inspection when necessary. Chapter 4 on compensation and damage authorizes the government quality supervision and control departments to order the producers to correct any violation (Article 28). Chapter 5 on penalty provisions states that a violating producer can be ordered to stop production, to pay a fine, to have its license revoked, or to take criminal responsibilities if appropriate (Article 38).

There are now established two administrations in the State Council to carry out the provisions of the Product Quality Law. The General Administration of Quality Supervision, Inspection and Quarantine of the PRC is responsible for certifying the quality-control systems of enterprises, issuing certification of product quality, inspecting possibly hazardous products, and enforcing all aspects of the Product Quality Law. The State Food and Drug Administration of the PRC is responsible for the approval of food and drug products for sale in the market. We need to collect more evidence to determine whether these two agencies are doing more good in protecting the consumers than harm in discouraging innovations. One important function served by certification of product quality is to facilitate the promotion of exports. A few examples of low-quality or mislabeled Chinese products in the world market may hurt the sale of other Chinese products. Some buyers of these products may desire government certification to ensure product quality. Even in this case of certification of quality of exports, readers of *Free to Choose* can present arguments in favor of a market solution.

The Law of the People's Republic of China on the Protection of the Rights and Interests of Consumers may have been partly inspired by the United Nations Guidelines for Consumer Protection adopted by the General Assembly on April 9, 1985. Chapter II lists as consumer rights the right to the safety of person in the purchase or use of a commodity; to knowledge of the true facts concerning commodities purchased; to require relevant information of a business operator providing commodities on price, place of origin, specification…; to free choice in purchasing commodities or services; to freely choose a business operator providing commodities or services; to compare, appraise, and select when freely choosing a commodity or service; to fair dealing; to fair terms of trade; and others. Chapter III specifies obligations of business operators and states, "A business operator providing a commodity or service to a consumer shall perform obligations in accordance with the Product Quality Law of the PRC and other relevant laws and regulations."

It seems that the consumer rights and the business obligations stated above are either unnecessary in the sense that the consumer obviously has such rights unless the government explicitly prohibits them or redundant in the sense that other laws already cover them.

Chapter VI on dispute resolution provides five ways for consumers to resolve disputes with business operators: (1) negotiate a settlement with the business operator; (2) request a consumer association to mediate; (3) complain to the relevant administrative department; (4) apply to an arbitral body for arbitration; or (5) institute legal proceedings in a People's Court.

For the purpose of this paper, a key issue is whether there is a large bureaucracy to enforce the consumer protection laws that may lead to an undue expansion of government power at the expense of economic innovations. To the extent that disputes are settled by other means than through an administrative department, the tendency of expansion of government power is restricted. In China, the population is accustomed to seeking help from the government, and cheating by business operators is common. Many consumers welcome intervention by the government to protect their interests from violation by other citizens. If they are free to choose, they may well choose an amount of government protection that is not very different from what is being practiced.

MACROECONOMIC POLICY

On monetary policy, the Chinese government agrees with the Friedman view that the quantity of money is the main instrument to control inflation, and it has applied this policy to maintain fairly stable price levels since the beginning of the PRC. However, there were periods in which the government failed to control the increase in money supply in order to stabilize the price level.

Before the establishment of the PRC, China experienced hyperinflation under the Nationalist government. The cause of hyperinflation was a large increase in the supply of currency as the government printed money to finance the civil war with the Communists and its other expenditures. Inflation created serious discontent and contributed to the downfall of the Nationalist government. Immediately after the establishment of the PRC, a new currency, renminbi (or the People's currency), was issued in exchange for the existing currency at a reasonable rate.

By control of the supply of renminbi, inflation was stopped within a few months. The general retail price index of China was relatively stable from 1952 to 1978, changing from 82.27 to 100.00. There was only one episode of inflation in the early 1960s, when the general retail price index increased from 93.08 in 1960 to 112.29 in 1962. We can interpret this increase by the quantity theory of money to which the Chinese government also subscribed as the quantity equation could be found in Marxian economic textbooks used in Chinese universities. Currency in circulation increased from 9.61 billion in 1961 to 12.57 billion in 1962, while the real GDP index (with 1978 = 100) reduced from 43.9 in 1960 to 30.9 in 1961. The drastic reduction of real GDP was the result of the failure of the Great Leap Forward movement introduced in 1958.

The control of the quantity of currency in circulation became less strict in the 1980s, when the government devoted more attention to the reform of economic institutions and allowed the expansion of credit to finance economic development activities. The first large monetary expansion occurred in 1984, when currency in circulation increased by 50 percent from January 1984 to January 1985, leading to an 8.8 percent inflation in 1985. Rapid monetary expansion at annual rates of over 20 percent continued in the following years until 1988, when the rate of increase was 48 percent and the corresponding inflation rate in the fall was about 30 percent annual rate. This inflation, together with government corruption, was a cause of the discontent and demonstrations in Tiananmen Square that ended in the tragic events of June 4, 1989.

After a slight and short-lived economic slowdown following the Tiananmen incident, the Chinese economy resumed its rapid growth at annual rates of over 10 percent in 1992 and 1993. The growth was stimulated by the announcement of Deng Xiaoping in Shenzhen in February 1992 to resume and even make bigger steps in reforming the domestic economy and opening China's door to foreign trade and investment. In 1992 money supply increased by 36 percent. Inflation as measured by the general retail price index reached 13 percent in 1993 and 22 percent in 1994.

It was Zhu Rongji, as the head of the People's Bank and later as vice premier and premier, who tightened the supply of money and credit and stabilized the price level. In fact, China's retail price index was reduced from a high of 121.7 in 1994 to 97.4 in 1998. Monetary policy in the period of the Asian financial crisis of 1997–99 might even have been too restrictive.

On fiscal policy, the Chinese government believes in using government expenditures to stimulate the economy during periods of slow growth. A notable example occurred in 1998 during the Asian financial crisis and a period of slower growth in China. Premier Zhu Rongji stated at a press conference on March 19, 1998, that to achieve an 8 percent growth rate in 1998, the main policy would be to increase domestic demand. He said that "to stimulate domestic demand, we will increase investment in construction of infrastructure, such as railways, highways, agricultural land and water conservancy facilities, municipal facilities, and environmental protection facilities. We will also increase investment in high-tech industries and in the technical renovation of existing enterprises." Thus, China adopted the Keynesian way of stimulating aggregate demand by increasing government expenditure, especially in infrastructure building.

In China, increasing government spending served not only to stimulate aggregated demand in periods of slow growth but to build infrastructure and other investment projects deemed necessary by the government. Government investment is regarded as important to achieve economic growth, which is not to be left entirely to the decisions of private investors in a market economy.

THE TIDE IS TURNING

I use the same title for this section as the title of the last chapter of *Free to Choose* because in both cases the subject is trends toward more freedom. In the last chapter of *Free to Choose*, the Friedmans suggest ways to increase freedom in the United States. Our discussion in this section is confined to reporting trends toward more freedom in China.

Following the seven topics discussed in this paper, consider first the general trend of economic freedom. We see that the trend is positive as market forces expand. The Chinese people are richer and can enjoy economic freedom to a larger extent. The government has acquired understanding of the working of the market economy and has encouraged entrepreneurship. More foreign competition in the domestic market and more opportunities to compete in the world market due to China's World Trade Organization membership are pushing the economy forward and will promote more economic freedom.

On political freedom, we can see progress toward a more democratic government coming from both the demand for and supply of democratic institutions. On the demand side, as the Chinese people have more economic power and become more educated, they will demand to have more freedom and more influence in governmental affairs. On the supply side, the Chinese Communist Party and government officials will become better informed of the modern political systems of the world. As they acquire confidence and ability to govern a modern society in the course of further economic progress, they will be more willing and able to adopt democratic institutions. The change toward a more democratic government has been observed in the spread of popular elections in Chinese villages, the increasingly independent behavior of members of the National People's Congress from directions of the Communist Party, and the improvement in the practice of the rule of law, partly as a result of China's need to deal with foreign corporations in international trade and investment.

A recent example that the Chinese government can improve from experience is its handling of the problem of severe acute respiratory syndrome (SARS).

On June 8, 2003, when the SARS problem had mostly subsided, *People's Daily* carried an article entitled "SARS, a valuable lesson for the Chinese government to learn." This article states that

> only by actively upholding the citizens' right to know can the government be better supervised by the public and in turn win the trust and respect of those it serves.
>
> People are made aware of government's views through the information it releases, and they exercise their rightful supervision not only through related government agencies but also through the media, which helps keep the government abreast of public opinion. Therefore, an interactive relationship among government, citizens and the media should be put in place so that the government knows the viewpoints of the people about its policies....The right afforded to the media and law to supervise should be fully guaranteed. When such a right is firmly in place, the activities of those in power come under public scrutiny, thus government and officials become publicly accountable for what they do and therefore more likely to work to higher standards.

The October 2, 2003, issue of the *New York Times* (A12) carries an article titled "China's Leader Calls for 'Democratic' Changes" and reports that President and General Secretary Hu Jintao, in an address to the governing politburo, said the Communist Party must undertake a "sweeping systemic project" to increase public participation in government and enforce the rule of law. "We must enrich the forms of democracy, make democratic procedures complete, expand citizen's orderly political participation, and ensure that the people can exercise democratic elections, democratic administration and democratic scrutiny."

This appears to be a sign of progress toward a more democratic government. Americans accustomed to a democratic government under a two-party system might find it difficult to appreciate a democratic government under a one-party system, but I believe that election of government officials and of members of the People's Congress is possible if the Communist Party is willing to put up for election the best candidates, who may be nonparty members, for the positions in question. Under a one-party system there are ways that citizens can participate in and influence political affairs. China might well turn out to be an innovator of one form of democratic government under a one-party system.

On the role of government, we observe that the Chinese government is playing its role in maintaining law and order, which is essential for the market economy to function. When the government is engaged heavily in economic activities, free choice of the people is provided in China by allowing competition among state enterprises, among state and private institutions, as in high school education, in health care delivery, in employment, and in the control of pension funds.

On social welfare, a new social security system is being put into effect that gives the people a wider choice than the previous system of entitlement to job security, retirement income, and health care, although the performance of the new system needs to be further studied. On education, we have observed the expansion of privately financed schools and the increase in tuition for students in higher education. Both are expected to continue.

On consumer protection, the People's Congress has enacted new laws on product quality and consumer rights along the lines that are set out by a 1984 resolution of the United Nations General Assembly. There is as yet no evidence of an unduly large bureaucracy for its administration that hampers innovations in consumer products.

Finally, on macroeconomic policies, the Chinese government recognizes the importance of monetary policy and has a fairly good record in controlling the supply of money to maintain a stable price level. At the same time, it practices Keynesian fiscal policy to simulate the economy and is active in building economic infrastructure and promoting new industries. If we allow for the possibility that the government of a developing country needs to play an active role in maintaining social and economic order, in fostering the development of market institutions and in promoting the development of some new industries, the record in providing freedom to choose in China has been reasonably good and is improving.

NOTES

The author acknowledges with thanks helpful comments from Zijun Wang on an early draft of this paper.

REFERENCES

Chow, Gregory C. 2002. *China's Economic Transformation*. Oxford: Blackwell Publishers.

———. 2004. *Knowing China*. Singapore: World Scientific Publishing Co.

Friedman, Milton, and Rose D. Friedman. 1980. *Free to Choose: A Personal Statement*. New York: Harcourt Brace Jovanovich.

Session 5

Financial Markets and Economic Freedom
Luigi Zingales

Choosing Freely: The Friedmans' Influence on Economic and Social Policy
Allan H. Meltzer

Friedman's Monetary Framework: Some Lessons
Ben S. Bernanke

Financial Markets and Economic Freedom

Luigi Zingales

The central point of Milton and Rose Friedman's best seller *Free to Choose* is that markets leave freedom of choice to individuals, while governments negate it. While this basic argument is clearly right, it is often objected to on grounds that it holds only for the well-offs. For many poor people around the world, the freedom markets grant seems to be more theoretical than real.

Consider, for instance, Sufiya Begum, a poor Bangladeshi stool maker. Theoretically, she is free to choose what to produce and for whom to produce. But she lacks the twenty-two cents necessary to buy the raw material for the stools she makes. Not having that money and lacking any opportunity to borrow it, Sufiya has no other choice but to accept the terms offered by the only middleman in town. He lends her the raw material but requires her to sell the stools back to him. Of course, he sets the price. Thus, lacking the initial resources, Sufiya is de facto enslaved to the middleman, who pays her only two cents for a hard day's labor.

One could quickly dismiss this argument on the basis of its assumption. Why is there only one middleman in town? Markets provide freedom of choice only when there is competition. Without competition, there is no freedom. Milton and Rose would certainly agree with that. Does this imply, however, that Milton and Rose's argument in favor of markets depends crucially on the feasibility of competitive markets? If so, its power might be significantly reduced. While Milton and Rose probably do not believe in the existence of natural monopolies, many economists do, especially when we are talking about poor developing countries. And regardless of the theoretical arguments, there are the facts. In Jobra, Sufiya Begum's village, there *was* only one middleman. So what freedom did markets offer to her?

We think instead that the major obstacle to Sufiya Begum's freedom of choice is not the existence of just one middleman in town but the lack of access to finance. If she could have access to some form of borrowing, Sufiya Begum could shop around for better prices and free herself.

If you had any doubt, consider the way access to finance has changed the life of another Bangladeshi woman, Delora Begum (no relationship with Sufiya). Delora was no different from Sufiya and millions of Bangladeshis, living in a straw hut with a corrugated metal sheet for a roof, a mud floor, and no toilet or running water. Thanks to a loan from the Grameen Bank (a development bank promoted by Bangladeshi economist Muhammad Yunus), however, she was able to acquire a Nokia cellular phone.[1] In a region of the world where it would be a compliment to call the traditional telephone network unreliable, the phone has made a world of difference in her life and in the lives of her fellow villagers.

The phone brings information at low cost to the farmers and tradesmen in the village; farmers can learn the fair value of their produce in the market, giving them an edge in bargaining with the notoriously exploitative middlemen; carpenters can find the current price of wood so they can get a better price for their furniture. The phone also reduces the cost of doing business. A local brick factory owner can order supplies on the phone rather than wasting time making the two-and-a-half-hour, bumpy, biweekly trip to Dhaka. Perhaps even more important, the phone helps save lives. A pregnant woman's life was saved when a call on the phone brought a doctor who could help.

The phone costs $375, and the monthly profits Delora Begum makes on it are about twice the average national monthly income. It has changed her life. Not only has she bought new possessions like a table and chairs, she now has status: "People consider me a person of honor," she told the *Wall Street Journal* staffer who reported the story.

While the cellular technology that frees her from the tyranny of the state-owned phone lines plays a big part in this story, the availability of finance plays an even bigger one. The availability of finance ensures the maximum individual economic freedom possible. In the words of Mohammed Yunus, the founder of the Grameen Bank:[2]

> If we imagine a world where every human being is a potential entrepreneur, we'll build a system to give everybody a chance to materialize his or her potential. The heavy wall between the "entrepreneur" and "labor" will be meaningless. If labor had access to capital, this world would be very different from what we have now.

Yunus' inspiring words nicely summarize the centrality of access to finance to economic freedom. Yet is there any systematic evidence to support his statement? This is what we turn to next.

FINANCE AND COMPETITION

Competition is what ensures our economic freedom. Competition gives us choices, improves the quality of the goods and services we buy, protects us against exploitation. Does greater access to finance promote competition?

Financial development seems to facilitate new entry. The deregulation of U.S. banking led to a substantial increase in the degree of development and competition in the financial sector in states that deregulated. These states experienced a significant increase in the rate of creation of new enterprises after deregulation.[3] Similarly, more new establishments are created in countries with more advanced financial systems, and this effect is more pronounced in industries that depend more on external finance, suggesting that availability of finance is indeed the cause of this increase.[4]

That financial development facilitates new entry, however, does not mean it necessarily reduces economic concentration: Established firms could benefit even more from financial development than entrants, acquiring a greater share of production and squeezing out competition. We can therefore look at the effects of financial development on competition more directly.

One measure of the degree of competition is the profit margin. All other things being equal, a firm in a more competitive market should have a lower profit margin. If firms in areas that have better access to finance have lower profit margins, this would suggest access to finance makes competition more intense. To check this, let us go to Italy, where there are tremendous variations across regions in the quality of the financial system. (Milton and Rose would not be surprised to learn that such difference was caused by a 1936 government regulation.)[5] As a result of these differences, an individual with similar personal characteristics has twice the probability of being rejected for a loan in certain Italian regions than in others, even after adjusting for economic factors that should matter.

These regional differences in access to finance seem to affect competition at the local level. Firms in the most financially developed regions have a profit margin 1.6 percentage points lower—about a third below the average profit margin of 5.9 percent—than in the least financially developed region. Reassuringly, this effect is present only for small and medium firms. Large firms can raise funds nationwide, and competition in industries with large firms should not be affected by local financial development.[6]

Perhaps the most reassuring evidence of the effects of finance on competition would come if we found an industry where access to capital is the primary barrier to entry and then compared competition in this industry across a number of countries over time.

Such an industry indeed exists—the cotton textile industry—and it has been studied by Stephen Haber.[7] The minimum economic scale of production in this industry has, historically, not been large; therefore, incumbent firms could not build tremendous barriers to entry by setting up massive plants. Moreover, over the period it was studied—approximately the second half of the nineteenth century and the first half of the twentieth century—there were few important patents in the industry and advertising did not play a major role. As a result, the main barrier to entry was the financing required to acquire the small but not insignificant amount of plant, machinery, and working capital needed for production. Haber compares the industry in two countries (Mexico and Brazil) at a similar level of economic development.

In 1883, Brazil had forty-four textile firms with approximately 66,000 spindles. The fraction of sales accounted for by the four largest firms was as high as 37 percent. In 1878, Mexico had almost twice as many firms, four times as many spindles, and its four largest firms accounted for only 16 percent of sales. Thus, the Mexican textile industry started out much bigger and less concentrated.[8]

In 1883, however, Brazil liberalized entry in its banking sector, while Mexico did not. A few decades later, the relative position in the textile industry was reversed. In 1915, Brazil had 180 firms, nearly 1.5 million spindles, and the four largest firms accounted for only 16 percent of sales. By contrast, in Mexico in 1912, there were only 100 firms, approximately three-quarters of a million spindles, and the four largest firms now accounted for 27 percent of sales.[9] Not only did Brazilian industry grow faster, it also became less concentrated!

FINANCE AND INDIVIDUAL MOBILITY

What we have just seen is that finance helps upstart corporations enter and challenge the establishment, thus keeping competition keen and refreshing. Let us now see if it also helps expand opportunity for individuals.

One obvious way to measure how finance expands opportunities is to assess its impact on the probability that individuals will start out on their own. Self-employment, whether as a plumber or a storekeeper, typically requires initial funds. For all but the lucky few who were born wealthy, the financial system is the only source for these funds. Thus, a more developed financial system should make it easier for individuals to become self-employed. What is the evidence?

Across Italian regions, differences in availability of finance translate into differences in individual mobility. Even after controlling for other regional differences, an individual living in the most financially developed region is 33 percent more likely to start out on her own than an individual with the same characteristics living in the least financially developed regions.[10] By reducing the importance of initial wealth, financial development also allows people to start out younger on their own. In the most financially developed regions, entrepreneurs are on average 5.5 years younger. Thus, financial development has a significant impact on economic mobility.

Another way to measure mobility is to look at the very rich and see how they came by their wealth. Since 1987, *Forbes* has put out a list of the world's richest people, indicating whether they inherited the bulk of their wealth or whether they are self-made. The last year for which *Forbes* reported all the people whose wealth exceeded $1 billion was 1996 (after that, the number of billionaires became too large). We start by looking at the number of billionaires present in each country in 1996. This is largely before the Internet boom created a whole new generation of instant (and ephemeral) billionaires. To compare countries with very different sizes, we divide the number of billionaires by how many million people live in that country.

The country with the highest frequency of billionaires per million inhabitants is Hong Kong (2.6), followed by Bahrain and Switzerland (1.7), and Singapore (1.4). At the bottom of the distribution we find poorer countries like Peru (0.04) but also rich countries like Norway (0) and South Africa (0.05).

More revealing than the frequency of billionaires is the frequency of *self-made* billionaires per million inhabitants. Not surprisingly, countries that have grown fast recently, such as Japan and the Asian Tigers—Hong Kong, Singapore, South Korea, and Taiwan—tend to have a high frequency of self-made billionaires. What is interesting is that the frequency of self-made billionaires per million inhabitants in the United States (0.26 per million) is much higher than that in the large European countries; the United Kingdom, Germany, and France have on average 0.08 self-made billionaires per million.

There is a very strong positive correlation between the frequency of self-made billionaires in a country and the size of its equity market. This correlation is not just due to Hong Kong, which stands out on both measures. If we eliminate the former British colony, the positive correlation persists. An increase in the size of the equity market from the level in France (50 percent of GDP in 1996) to the level in the United States (140 percent of GDP) would be associated with an increase in the frequency of self-made billionaires in France from 0.07 per million to approximately 0.3. All the difference between the United States and France in the frequency of self-made billionaires per million inhabitants can be explained by the better-developed financial markets in the United States![11]

All we have is a correlation, which, as we have previously emphasized, is not evidence of causation. A lot of other factors, such as the extent to which a country favors free enterprise, may affect the ability of an individual to accumulate a fortune during her lifetime. As long as these other factors are relatively constant, however, we can eliminate their influence by looking at changes in the frequency of billionaires per million people over a certain period of time. Since the earliest survey conducted by *Forbes* is in 1987, we look at the changes in the frequency of self-made billionaires between 1987 and 1996. During this decade, the frequency of self-made billionaires tends to increase most in countries that started the decade with a more developed financial system. The effect on the frequency of inheriting billionaires is much smaller.

THE INCREASED VALUE OF HUMAN CAPITAL

Until around the middle of the nineteenth century, the U.S. (and world) economy had few firms with more than a hundred employees. Most were managed by their owners. In the latter half of the nineteenth century, a new organizational form emerged: large, vertically integrated firms that Alfred Chandler calls the *modern business enterprise*. One of the key characteristics of this new form of organization was its capital intensity. Between 1869 and 1899, capital invested per worker in the United States nearly tripled in constant dollars.

This capital intensity coupled with the underdeveloped stage of finance gave these firms an enormous staying power. Of those firms on the U.S. Fortune 500 in 1994, 247, or nearly half, were founded between 1880 and 1930. The early firms include Kodak, Johnson & Johnson, Coca-Cola, and Sears, all founded in the 1880s, and General Electric, PepsiCo, and Goodyear in the 1890s.[12] Firms have been even more durable in Germany. Of the thirty largest German firms ranked by sales in 1994, nineteen were founded between 1860 and 1930 and four even earlier. The nineteen firms include household names like Daimler Benz, BMW, Hoechst, Bayer, and BASF.[13]

As a result, workers, even very senior managers, could not contemplate life outside the firm. The only source of protection was coming from competition from other firms, not from the possibility of breaking away and creating their own firm. Competition works best for more homogenous products, so unskilled workers were relatively better protected than more senior managers, who had more firm-specific human capital. Consistently, between 1890 and 1950—the period of the rise of Chandler's large industrial enterprise—there was a tremendous compression of the wages of educated white-collar workers relative to blue-collar workers.[14] The ratio of wages of clerical employees to those of production workers fell from approximately 1.7 to 1.1 between 1890 and 1950. Because typically the educated are also relatively more skilled, these facts are consistent with the consolidation of industry into large, monolithic organizations shackling the skilled and compressing the wage differential. Of course, other factors also partly account for this compression. As with all economic resources, demand and supply ultimately determine the relative prices of skilled labor. Our point is simply that access to finance may have had a profound influence on both demand and supply.

Developments in finance in the last three decades, however, have inverted this trend. With the diffusion of leasing arrangements and highly leveraged transactions, alienable assets such as plant and equipment have become less unique, especially to those with specific skills. This has reduced the barrier to entry created by high capital intensity. As a result, managers in a division are no longer beholden to the parent because the latter owns their assets. If need be, they can break away, raise finance directly in the market, and replicate the assets. The possibility of starting up on one's own has opened up new possibilities for those who, hitherto, were locked into their firms because of specialization—in particular, the older workers whose mobility seems to have increased. From the firm's perspective, resources other than alienable assets have become more critical to its ability to survive competition. From the owner's perspective, these resources—people, ideas, strategies—are harder to control directly. Because specific skills are not only more valuable in this era but also more mobile, it is no wonder that skilled workers are being paid more. This rise in competition has put a premium on skills. Consistently, between 1979 and 1987, the average weekly wage of college graduates with one to five years of experience increased by 30 percent relative to the average weekly earnings of comparable high school graduates.[15]

Even more important, the options created by the availability of finance have altered the balance of power between capital and labor, in favor of the latter. The single biggest challenge for owners and top management today is to manage in an atmosphere of diminished authority. Authority has to be gained by persuading lower managers and workers that the workplace is an attractive one and one that they would hate to lose. To do this, top management has to ensure that work is enriching, that responsibilities are handed down, and that rich bonds develop between workers and between themselves and workers. The emphasis on a kinder, gentler firm in most recent management tomes is not without foundation.

Finally, in addition to increasing the wages of the skilled and increasing their power within firms, these changes have also increased the opportunities for innovation. Inventors in high-tech firms who do not feel their ideas are appreciated or rewarded can leave and get them financed elsewhere. Intel—one of the most innovative and profitable firms of our time—was started by Gordon Moore and Robert Noyce because they felt their employer, Fairchild Semiconductor, was ignoring important new technologies they believed in. More generally, a recent study reports that 71 percent of the firms included in the U.S. Inc. 500 (a list of young, fast-growing firms) were founded by people who replicated or modified an idea encountered in their previous employment.[16]

In a sense, therefore, improvement in finance has ensured the maximum individual economic freedom possible.

WHY WE DO NOT SEE MORE FINANCE

Given all the benefits finance brings to people, why don't we see more of it? Why between 1880 and 1915 did Mexicans have to suffer more than Brazilians? Why does access to finance differ so much across Italian regions? Why did it take Mohammed Yunus to give Delora what Sufiya does not have?

The Roadblock to Finance

As Raghuram Rajan and I elaborate in our book *Saving Capitalism from the Capitalists*, the answer is very simple. Financing is a very risky and complex activity. It implies exchanging a sum of money today for a promise of a bigger sum in the future. This promise is fraught with risks: the risk that people borrow with the intention of not repaying (adverse selection), the risk that they might not work as hard in repaying it (moral hazard), and the risk that the government intervenes, reducing the value of that promise. While the first two types of risks seem greater, it is the third one that creates the main obstacle to the development of finance. In fact, human ingenuity has created a lot of mechanisms to reduce adverse selection and moral hazard.

Consider, for instance, the tremendous challenge of lending in Bangladesh. Given the creaky pace of the legal system, it is all too easy for a borrower to take a loan, consume the proceeds, and then default. Hence, no honest banker is willing to lend unless the borrower is already trusted or wealthy. Corrupt bankers are willing to lend, of course, but the borrower has to kick back so much of the loaned amount to the banker for him to ignore credit standards that only borrowers who do not intend to pay loans back take them. Ordinary decent people like Sufiya Begum have to rely on local moneylenders who levy usurious rates of interest and use extralegal threats of violence to extort repayment.

Even in this extreme situation, human ingenuity has created an alternative. The Grameen Bank, the institution that lent money to Delora Begum, works on the same principle as the moneylender—the borrower should have an incentive to repay—but is much more civilized in its methods. Here is how it operates: A woman who wants to borrow (the Grameen Bank finds women more reliable) must find four other friends who are eligible for membership. The women cannot be related to each other, and they must agree to help each other when in difficulty. If any member of the group cannot repay, the others must help repay the defaulter's loan or risk seeing their own credit lines diminished. Not all the group get loans at the same time. Two members get a loan, and when they have established a record of prompt payment, the next two become eligible, and eventually the last member. As the group repays, it becomes eligible for larger and larger loans.

The beauty of this method is that initially, peer pressure from members within the group—who are often in the same social circles—forces repayment. Because the initial loans are for small amounts, peer pressure works. Moreover, there is group support in the case of untoward accidents, which occur all too often when one is living at the margin. But eventually, the group's excellent record of past repayment becomes an asset—its reputation—that is worth preserving in its own right because it is the source of access to credit. While the bank has to coach the group intensively at high cost for the first few loans, eventually the group becomes as creditworthy as any rich rice merchant in Dhaka and graduates to larger, more cost-effective loans. Repayment rates on loans to even the very poor are extremely high, allowing the bank to charge moderate rates of interest.

So creative arrangements can overcome many of the challenges posed by moral hazard and adverse selection. They find it impossible, however, to overcome the interference of the government. Consider, for instance, the case of another developing country with poor law enforcement: Paraguay.[17] In that country, the major obstacle to lending is not due to the absence of proper laws but to the presence of crazy ones. One law, for instance, allows creditors who lend small amounts (up to $2,650) to take as collateral only small assets such as appliances and prevents them from collecting more than 25 percent of the value of these assets. To make the problem worse, the lack of public registry allows borrowers to trade the goods posted as collateral without the lender's knowledge. So the collateral often disappears by the time the creditor realizes the borrower intends to default. The result is that creditors are reluctant to make loans even if the borrower has the collateral to pledge. While this law was sold as a way to protect the poor from the rapacity of the financiers, it ended up damaging the very people it allegedly wanted to help.

Paraguayans' ingenuity, however, found a way around these restrictions by using postdated checks. Until 1996 checks written against accounts without funds in them (bounced checks) were considered a form of swindling, and those who wrote such checks could go to jail. So debtors who had no collateral could sign a check for the principal amount to be repaid with the future date on which the loan matured on it. If on that date the borrower did not have the money to pay, lenders could go to the bank, present the check, and have payment formally refused by the bank because of inadequate funds in the account. They could then take the bounced check to a judge, who would issue an order of arrest. If the debtor still did not pay, the lender could put him in jail. This threat ensured that borrowers would pay up if they could because they would do anything to avoid going to prison.

The outcome if the lender defaults is, of course, not civilized. It harkens back to Dickensian times and the debtors' prisons that Mr. Micawber frequented. And defaults are not infrequent, suggesting that many on the margin are willing to take the risk of incarceration just to obtain financing. One estimate of the criminal proceedings in Paraguay that relate to bounced checks is between 30 and 40 percent. In the 1990s, about 10 percent of the National Penitentiary population was imprisoned because of checks issued without funds. That so many were willing to take the risk of losing their freedom to obtain financing suggests how important it is in ameliorating their lives.

In 1996, however, the Paraguayan parliament eliminated the criminal sanction on bounced checks. While this might appear as a very humanitarian measure meant to reduce the number of inmates, it had devastating effects on the credit market. Without the criminal penalty, entrepreneurs could not post their future livelihood as collateral. But because the other law, limiting the ability to collateralize assets, was not repealed, they had no collateral to post. It is no surprise that the volume of lending dropped.

At first sight, the perverse effect of the repeal of one law and the upholding of the other could be perceived as an innocent mistake of a well-meaning Paraguayan government. Unfortunately, these "mistakes" are so common and so pervasive, even in more developed countries, that it is hard to believe in sheer bad luck.

For example, in an attempt to protect households from the consumer credit industry that "forced" them to take too much debt, the Commission on the Bankruptcy Laws of the United States argued in 1973 that it would be beneficial for less-well-off households if they could retain some assets after filing for bankruptcy. Following these recommendations, a number of states adopted exemptions. Some of these were extremely generous. For instance, in Texas, a bankrupt person can retain his house no matter how expensive it is, in addition to $30,000 of other property. A bankruptcy exemption is a form of insurance—it prevents the borrower from losing everything in case of a personal calamity. This can make borrowers more willing to tolerate high debt levels. But it also prevents the exempt assets from serving as collateral, making lenders less willing to offer loans. Not surprisingly, higher state bankruptcy exemptions led to a significantly higher probability that households would be turned down for credit or discouraged from borrowing.[18]

More interestingly, poor households were disproportionately adversely affected. Because their house is often their only form of collateral, the exemption laws effectively deprived them of their only means of obtaining finance. They had much less access to borrowing in the high-exemption states, and paid higher interest rates, than in the low-exemption states. By contrast, rich households typically have enough unprotected assets to borrow. The diminished willingness of financiers to lend after the passage of the exemptions did not affect them much. In fact, their debt went up in high-exemption states: Rich households became more willing to borrow because more of their assets could be protected from seizure. Thus, financial legislation that was intended to help poor households ended up hurting them and benefiting the well-to-do. These examples are more common than one could ascribe purely to legislative ignorance of economics.

Cui Prodest?

In every good mystery novel, when coincidences become overwhelming, the detective starts looking for a motive. Why do governments throughout the world, with the pretense of helping the poor, hurt them by depriving them of access to finance? Who benefits from it?

If finance does indeed breed competition and favors social mobility—as we argue it does—the people who will lose from financial development are those currently well-off, the owners of the established firms, who fear competition and would like to freeze societal and economic positions at their current level.

Consider, for instance, established large industrial firms in an economy, a group we will call industrial incumbents. In normal times, these incumbents do not require a developed financial system. They can finance new projects out of earnings—as most established firms do—without accessing external capital markets. Even when their business does not generate sufficient cash to fund desired investments, they can use the collateral from existing projects and their prior reputation to borrow. Such borrowing does not require much sophistication from the financial system—even a primitive system will provide funds willingly against collateral. Because of their privileged access to finance in underdeveloped financial systems, incumbents enjoy a positional rent. Anybody else who starts a promising business has to sell it to the incumbents or get them to fund it. Thus, not only do incumbents enjoy some rents in the markets they operate in, but they also end up appropriating most of the returns from new ventures.

These rents will be impaired by financial development. The better disclosure rules and enforcement in a developed financial market will reduce the relative importance of incumbents' collateral and reputation, while permitting newcomers to enter and compete away profits.

Similar arguments apply to incumbent financiers. While financial development provides them with an opportunity to expand their activities, it also strikes at their very source of comparative advantage. In the absence of good disclosure and proper enforcement, financing is typically relationship-based. The financier uses his connections to obtain information to monitor his loans and uses his various informal levers of power to cajole repayment. Key, therefore, to his ability to lend are his relationships with those who have influence over the firm (managers, other lenders, suppliers, politicians, etc.) and his ability to monopolize the provision of finance to a client (either through a monopoly over firm-specific information or through a friendly cartel among financiers). Disclosure and impartial enforcement tend to level the playing field and reduce barriers to entry into the financial sector. The incumbent financier's old skills become redundant, while new ones of credit evaluation and risk management become necessary. Financial development not only introduces competition, which destroys the financial institution's rents and its relationships, it also destroys the financier's human capital.

After the 1994 Mexican crisis, for instance, the World Bank decided to help the government improve the financial infrastructure. One of the fundamental institutions that were missing was a credit registry, where assets posted as collateral for a loan could be officially recorded so that any potential lenders could be aware of what a borrower had already pledged. In setting up this registry, the World Bank experienced strong resistance from the existing banks. Why? Existing banks had enough clout that they could get this information regardless of the credit registry. Not only would they not benefit from it, but they would see their competitive position eroded as less established lenders could access that information and compete for business on an equal footing. Access to credit was curtailed to support the interests of the few!

In sum, a more efficient financial system facilitates entry and thus leads to lower profits for incumbent firms and financial institutions. From the perspective of incumbents, the competition-enhancing effects of financial development may offset the other undoubted benefits that financial development brings. Moreover, markets tend to be democratic, and they particularly jeopardize ways of doing business that rely on unequal access. Thus, not only are incumbents likely to benefit less from financial development, they might actually lose. This would imply that incumbents might collectively have a vested interest in preventing financial development and might be small enough to organize successfully against it.[19] In doing so, they will be able to rely on other incumbent groups such as organized labor that previous studies have shown benefit from an economy with limited competition.[20]

The Spreading of Financial Markets

In spite of this strong opposition to finance, a veritable revolution has taken place in this field throughout much of the developed world in the last three decades. In 1970, the ratio of the value of all listed U.S. stocks to GDP was 0.66; by the year 2000 it had climbed to 1.5.[21] The increase in other countries is even more dramatic. In France, stock market capitalization rose from just 0.16 of GDP in 1970 to 1.1 times GDP in the year 2000.[22] The explosion in the size of stock markets is just one indication of what has happened. Entire new markets such as Nasdaq have emerged, catering specifically to young firms. Institutions such as money market funds did not exist in the early 1970s; now they hold over $2 trillion in assets in the United States. A large number of financial derivatives that are commonplace today, such as index options or interest rate swaps, had not been invented three decades ago. The turnover in the trading of such derivative instruments was $163 *trillion* in the fourth quarter of 2001, about 16 times the annual GDP of the United States.[23]

In the same way as corporations have obtained new instruments with which to raise finance and allocate risks, individuals also now have expanded choices. Revolving consumer credit such as credit card debt exploded from near nothing in the United States in the late 1960s to nearly $700 billion in late 2001.[24] Firms and individuals can borrow not just from domestic institutions but also from foreign markets and institutions. Gross cross-border capital flows as a fraction of GDP have increased nearly tenfold in developed countries since 1970 and more than fivefold for developing countries. In the decade of the 1990s alone, these flows more than quadrupled for developed countries.[25]

Why have politicians in countries as diverse as France and Germany or Korea and India embraced the market and attempted to provide the governance markets need? It is difficult to imagine that politicians have suddenly become more public-spirited. The answer, we believe, is that the interests of the elites have changed with the opening of borders to goods and capital. This has made domestic elites press their politicians to enact market-friendly legislation.

The effect of open borders on government policy can be clearly seen in the Indian software industry. Unlike much of Indian industry, which for years had been held back by government regulation intended to protect the positions of a few existing family-owned firms, there were no incumbents in this sunrise industry. Smuggling provided an initial impetus as cheap computer parts were brought in, bypassing extortionate tariffs and giving software engineers access to cheap computers. But as the industry grew, it gained political clout. Tariffs were brought down so the industry could work more legitimately. The industry also became internationally competitive.

Now, even when some players in the industry have become dominant, they are not particularly interested in suppressing domestic competition. Instead, people like Narayan Murthy, the head of Indian software giant Infosys, are pressing for better infrastructure, such as better corporate governance that will let them raise finance more cheaply and more educational institutions that will allow them to train their personnel. Everyone, including their domestic rivals, benefits. The software industry is pushing India toward building infrastructure for the market. Openness has been critical to this process.

In sum, as borders open up to the flow of goods and capital, incumbent firms now need well-functioning domestic markets so that they can take advantage of the opportunities provided by the global market, as well as meet foreign competition head-on. The prospect of increased domestic competition matters less when they are fighting on the world stage. They now back rather than oppose domestic markets. Put differently, competition between economies through open borders forces politicians to enact the rules that will make their economies competitive. This typically means enacting market-friendly legislation and making markets accessible to all.

CONCLUSIONS

Widespread access to finance is an integral pillar of our freedom to choose. Without it, many of the opportunities offered by the market are more hypothetical than real. In spite of the enormous challenges intrinsic to the financing activity, human ingenuity, when allowed to work freely, is able to devise many mechanisms to enlarge access to finance. It cannot, however, overcome the power of the government, when this is determined to block finance. Unfortunately, governments are too often captured by rich incumbents, who stand to gain very little and risk a lot from the development of finance.

Openness is a powerful antidote against the power of incumbents. International competition weakens both the incentives of incumbent firms to oppose financial development and their ability to do so. Not coincidentally, the last twenty-five years, which have witnessed an opening up of world trade, have also experienced a surge in financial development.

Unfortunately, international tensions and recent economic downturns have slowed down and threatened to reverse the process of world market integration. As economists, as disciples of Milton Friedman, but most important, as citizens, we have to fight this trend, which can significantly affect the freedom to choose of generations to come.

NOTES

[1] The description of Delora Begum's cell phone and its effects is from a *Wall Street Journal* article by Miriam Jordan, published in the June 25, 1999, edition and entitled "It Takes a Cell Phone: A New Nokia Transforms a Village in Bangladesh."

[2] Alex Counts, 1998, *Give Us Credit: How Muhammad Yunus's Microlending Revolution Is Empowering Women from Bangladesh to Chicago*, Random House Value Publishing.

[3] Sandra Black and Philip Strahan, 2002, "Entrepreneurship and Bank Structure," *Journal of Finance*, December.

[4] Raghuram Rajan and Luigi Zingales, 1998, "Financial Dependence and Growth," *American Economic Review* 88, Issue 3: 559–86.

[5] See L. Guiso, Paola Sapienza, and Luigi Zingales, 2003, "The Costs of Banking Regulation."

[6] See L. Guiso, Paola Sapienza, and Luigi Zingales, 2002, "Does Local Financial Development Matter?" NBER Working Paper No. 8923.

[7] Much of what follows draws on the work of Stephen Haber, an economic historian at Stanford University.

[8] Stephen Haber, ed., 1997, *How Latin America Fell Behind: Essays in the Economic Histories of Brazil and Mexico, 1800–1914*, Stanford University Press, 162–63.

[9] Ibid.

[10] This effect is present even controlling for other regional differences in economic conditions. See

L. Guiso, Paola Sapienza, and Luigi Zingales, 2002, "Does Local Financial Development Matter?" NBER Working Paper No. 8923.

[11] The pattern persists after we account for the influence of a country's per capita GDP and its recent growth rate. One obvious explanation is that billionaires are more likely to own publicly traded firms, and their stock prices tend to be high when the country's equity markets are high. Those who inherit wealth, however, also tend to own stock—the heirs of Sam Walton, the Fords, and the Siemens come to mind—but the relation between frequency of inherited billionaires per million people and stock market capitalization is much weaker. More important, the relationship exists even if we measure financial development in a way that is not directly affected by the level of stock market valuations—such as the number of listed firms per million of population. Countries with better accounting standards also have more self-made billionaires.

[12] See Harris Corp., 1996, "Founding Dates of the 1994 Fortune 500 U.S. Companies," *Business History Review*, Spring: 69–90.

[13] See J. Fear, 1997, "German Capitalism," in *Creating Modern Capitalism: How Entrepreneurs, Companies, and Countries Triumphed in Three Industrial Revolutions*, ed. T. McCraw, Cambridge: Harvard University Press, 181.

[14] Claudia Goldin and Lawrence Katz, 1999, "The Returns to Skill in the United States Across the Twentieth Century," NBER Working Paper No. 7126.

[15] L. Katz and K. Murphy, 1992, "Changes in Relative Wages, 1963–1987: Supply and Demand Factors," *Quarterly Journal of Economics* 107: 33–78.

[16] A.V. Bhide, 2000, *Origin and Evolution of New Business*, New York: Oxford University Press, 94.

[17] Our account is based on Stephanie Straub and Horacio Sosa, 2001, "Ensuring Willingness to Repay in Paraguay," in *Defusing Default: Incentives and Institutions*, ed. Marco Pagano, Inter-American Development Bank.

[18] R. Gropp, J. Scholz, and M. White, 1997, "Personal Bankruptcy and Credit Supply and Demand," *Quarterly Journal of Economics* 112: 217–51.

[19] For example, see Mancur Olson, 1965, *The Logic of Collective Action: Public Good and the Theory of Groups*, Cambridge: Harvard University Press; and George Stigler, 1971, "Theory of Economic Regulation," *Bell Journal of Economics* 2: 3–21.

[20] For evidence that unions share in rents from industrial concentration, see, for example, Michael A. Salinger, 1984, "Tobin's q, Unionization, and the Concentration–Profits Relationship," *Rand Journal of Economics* 15: 159–70; and Nancy L. Rose, 1987, "Labor Rent-Sharing and Regulation: Evidence from the Trucking Industry," *Journal of Political Economy* 95: 1146–78.

[21] For year 1970, R. Rajan and L. Zingales, "The Great Reversals: The Politics of Financial Development in the 20th Century," forthcoming in the *Journal of Financial Economics*, and for year 2000, data are from International Federation of Stock Exchanges.

[22] Ibid.

[23] Bank for International Settlements *Quarterly Review: International Banking and Financial Market Developments*, March 2002.

[24] Economic Report of the President, February 2002, 412.

[25] Ibid., 261.

Choosing Freely:
The Friedmans' Influence
on Economic and Social Policy

Allan H. Meltzer

At Milton Friedman's sixtieth birthday conference, in 1972, George Stigler, the dinner speaker and Milton's friend and colleague, discussed economists' or academics' influence, particularly Milton Friedman's influence. As I recall the lecture after more than thirty years, Stigler began by noting that Milton Friedman was among the most influential of all economists. Then he asked how influential that might be. The implicit answer was that maximizing individuals recognized their self-interest and acted accordingly. There was limited room for influence or persuasion. Influence had, at most, the modest role of hastening the adoption of better solutions.

To bolster his argument, Stigler chose the role of economists in repealing the British Corn Laws. With his typical irony, he drove home the point: "How heartening a tale! Economists turned a great nation from error to truth, from inefficiency to maximum output" (Stigler 1975, as quoted in Schwartz, 1993, 207). Stigler then gave his own explanation of the repeal of the Corn Laws. Economists' defense of free trade had a modest role.

I found this argument unpersuasive and, in fact, more than a little strange. It was made by an academic, a person whose life was devoted to teaching and research. These activities are useful only if there are ideas that are not known to students or not yet discovered or facts that are misperceived or misinterpreted. It was written by a superb essayist who spent much of his life trying to persuade, even influence, others. And it was based on a model of the political process that denigrated the role of ideas, particularly new ideas, in political campaigns.

Stigler's comments went unchallenged for two decades. On Milton Friedman's eightieth birthday, Anna Schwartz took the other side. She argued that Stigler's alternative explanation of the repeal of the Corn Laws was not compelling. She went on to cite some examples of Friedman's influence: the volunteer army, education vouchers, and repeal of interest rate ceilings. But she also cited some examples where his proposals had not been adopted; minimum wage laws and import quotas are two examples. Schwartz also included a flat-rate income tax with no deductions. Since that time, the U.S. tax schedule has become flatter, and the Russians and some other former communist countries have adopted a flat-rate income tax.

Not only is influence now an established theme when we celebrate the contributions the Friedmans have made alone and together, but I believe there is more to be said about influence. The Friedmans' efforts to change major aspects of society represented by *Free to Choose, Capitalism and Freedom* and many other works contain numerous suggestions and proposals that were adopted, some that were adopted in part or in modified form, and some that remain dormant and rarely discussed. I will suggest some reasons for these successes and failures.

Further, though I disagree with the main thrust of George Stigler's comments about influence, he raises an important question. In his terms, if economists convinced the British to repeal the Corn Laws and move decisively toward free trade, why did the same logic not persuade other governments at the time or in the next century? Why did it take seventy years for Adam Smith's argument about the benefits of free trade to be accepted in Britain? Stigler suggested that external conditions, particularly demographics, were the dominant influence. More generally, how important are external conditions, and how do they interact with the ideas of economists, social scientists, or reformers?

One of the benefits of writing this paper was that, to prepare it, I reread both *Capitalism and Freedom* and *Free to Choose*. These books are, in different ways, rich in proposals for changes that increase liberty and opportunity and suggestions about why some proposals might not be adopted. The main reasons given for expecting proposals to fail are bureaucratic inertia and myopia. Although there are references to theories of public choice, the authors mostly do not emphasize rational, maximizing public officials. An exception is the discussion of drug licensing (Friedman and Friedman 1980, 209).

The Friedmans conclude *Free to Choose* with a chapter that has a hopeful and even optimistic title: "The Tide Is Turning," a declarative statement, not a question. I will do the same, concluding by offering some thoughts on that subject twenty-five years after the publication of their book.

THE CLIMATE OF OPINION

Anyone under sixty years old may find it difficult to appreciate what the climate of opinion was in the 1930s and how much it has changed both within the economics profession and without. The dominant view then was that capitalism had failed; the future was some form of socialism, and the only issue was how extensive it should be. Keynes wanted free markets for consumer goods but state planning and direction of investment. Alvin Hansen claimed that market economies faced stagnation, unless the state managed investment to maintain full employment. There were opposing views. Schumpeter (1942) dismissed Hansen's argument about stagnation, but he, too, for very different reasons, saw socialism as the future.

Microeconomists discussed the "wastes of competition," for example, two or three milk companies delivering milk to the same streets. Planning could deliver the milk and avoid the waste. Oxford studies showed that businessmen never considered interest rates when making investments. Prices had little to do with resource allocation. George Stigler mocked these early econometric studies, offering Stigler's law: All demand curves are price inelastic.

Price theory was taught as mainly an exercise in applied geometry or, for the more advanced, algebra. Economic textbooks of that period, and even much later, offered few applications to problems. Agriculture was the exception. Price theory could show the effects of agricultural price supports. More adventurous authors used price theory to show the incidence of taxes. There was little discussion in economics of prices directing resource use or of crime, health care, education, and many other topics that are the daily concerns of modern economists. The marginal productivity theory of labor was taught but dismissed as lacking empirical content, an empty box.

In recalling that era, I don't think I have exaggerated. As always, there were exceptions, just as there are now Marxists, neo-Marxists and socialists of various types in economics departments teaching something other than price theory and its applications.

To me, personally, Hayek's *The Road to Serfdom* came as a shock. Later came *Capitalism and Freedom*, a book that presented as its major theme that the "organization of the bulk of economic activity through private enterprise operating in a free market promotes economic welfare and political freedom" (Friedman 1962, 4). By that time, I was a practicing economist and, partly under the influence of Armen Alchian and Karl Brunner, had given up my earlier leftist orientation. The numerous, creative applications in *Capitalism and Freedom* were then, as they are now, a treat to read and think about.

SUCCESSES AND FAILURES

The Friedmans, Milton especially, have had an enormous influence not only on economists and the academic profession but on policies in the United States and large parts of the world. Their efforts to induce societies to foster liberty, individual initiative, and freedom to choose and their successes have few parallels. One thinks of Smith or Marx, both of whom are still read and whose works are invoked as a basis for changes, albeit very different changes. Of course, there is Keynes the polemicist who wrote *The Economic Consequences of the Peace*, the economist whose *General Theory* changed economic theory and policy, and the social reformer whose *Essays in Persuasion* offered many proposals for social and economic change. Whether we favor or oppose the recommendations, this is distinguished company. Each of these economists continues to influence policies and interpretations of events.

My Oxford dictionary lists several definitions of influence. The relevant one refers to the power indirectly to affect the course of events. The only change I would make is to insert "directly and" before indirectly, so that the definition refers to direct and indirect influence on events. Friedman's influence on the military draft was direct. As I discuss below, he and others convinced the military and other officials to try a volunteer army.

By my count, there are more than twenty-five specific recommendations in *Capitalism and Freedom*, some additions, extensions, and repetitions in *Free to Choose*, and other proposals scattered through Milton Friedman's published works. I find it useful to divide the proposals, first, into those to which the Friedmans devoted considerable effort and those that received less of their attention, perhaps because implementation seemed unlikely. Two caveats apply. First, decisions about success and failure and about effort unavoidably have a subjective element. Second, some of the proposals may have been adopted in other countries. I offer some examples, but I do not have enough knowledge of practices throughout the world to claim accuracy.

Failures

Here are some proposals that have not been adopted and, I believe, are not currently under consideration. Each of these can be found in *Capitalism and Freedom*. Friedman proposed to abolish state universities; abolish licensing of doctors, lawyers, accountants, and other professions; eliminate agricultural subsidies, minimum wage laws, right to work and fair employment practice laws; and adopt a spending limitation amendment. In *Free to Choose* and elsewhere, the Friedmans amplified the last proposal and, with others, developed a proposed constitutional amendment to limit growth of government spending.

None of this has happened, and none of these proposals seems likely to be adopted soon, if ever. Of course, the future is not knowable, but none of these proposals is under active consideration currently. Some issues have moved farther from free choice.

About fifteen or twenty years ago, more than thirty states approved a call for a constitutional convention to adopt a spending limitation amendment to the Constitution. The Constitution requires thirty-four states to adopt the call. I followed the effort closely because one of our sons worked for the National Taxpayers Union, a group that sponsored the amendment. At the time, one of his responsibilities was to work with legislators to get the call for a convention approved. The effort failed in the states with relatively more unionized workforces. The early successes may have benefited from some free riders—legislators who voted aye because the number of states remained below the constitutional requirement. I count that as a failure.

In *Free to Choose*, the Friedmans proposed to phase out Old Age and Survivors Insurance (OASI). They would honor existing obligations, repeal the payroll tax, and rely on voluntary decisions about pensions. They predicted that their proposal "has no chance whatsoever of being enacted at present" (1980, 124). Stigler's comment about demographics seems apposite. Recently, President Bush proposed private management of a part of OASI accounts, a small but important step in the direction the Friedmans proposed. The incentives for the currently younger generation to seek to increase their return on pension assets suggests that the "at present" in the quotation is more critical than at first appears. The Friedmans' proposal has not been adopted, and it seems unlikely that the government will withdraw completely from managing pensions and redistributing income intergenerationally. A partial success seems more likely now than when they wrote.

Complete Successes

I count four complete successes—proposals that became law or policy without major change from the Friedmans' proposals. Three of the four are of considerable importance. Each is a shift from command and control regulation to free markets and free choice. The fourth success, the right of U.S. citizens to own, buy, and sell gold, I regard as less important because there are many alternative assets and instruments. Fortunately, the time has never come when ownership of gold was needed here to protect wealth from a tyrant.

The three major successes are floating the dollar, ending the military draft, and removing interest rate ceilings on demand and time deposits. The three have in common that each change occurred in response to a crisis; the existing system failed. In each case, there was at least one alternative solution that prevented free choice in competition with the Friedmans' proposals. One lesson these experiences illustrate is that a market solution is not a government's first choice, or the obvious alternative in a crisis. Free market solutions have a greater chance of success if the proposals are known in advance, proponents have responded to criticisms and objections, and officials have become sufficiently familiar with the proposal that they believe it can work.

Floating the Dollar. Floating the dollar in 1971 and 1973 illustrates these principles. Milton Friedman first proposed floating exchange rates in the early 1950s. At the time and for many years, there was no interest among politicians. Friedman (1953) predicted that the Bretton Woods system of fixed but adjustable exchange rates would break down. By the 1960s others began to recognize that the system was in trouble. The stock phrase at the time was that there were three problems: liquidity, adjustment mechanisms, and confidence. Though repeated endlessly, the main effort went to create additional liquidity, the SDR. Governments and their officials solved a problem that did not exist and ignored the overvaluation of the dollar. There were ample reserves of dollars available, and the supply continued to grow. Except for France, foreign governments did not wish to abandon the dollar or the fixed exchange rate system. They wanted the United States to do the impossible—reduce the supply of dollars without deflating or reducing U.S. imports. With the modest exceptions of Germany, the Netherlands, Switzerland, and Austria, countries did not revalue their currencies, and only France demanded devaluation of the dollar against gold.

A personal anecdote illustrates the state of informed discussion. Many meetings and symposia on the dollar and the international monetary system considered proposals for monetary reform. At one in the summer of 1968, organized by Edward Bernstein and hosted by David Rockefeller, several academics met with prominent bankers and government officials. The agenda included proposals for the SDR and return to a gold standard. Floating exchange rates was not on the program.

Gottfried Haberler and I proposed to Bernstein that this should be added. After some discussion with the hosts, he told us that floating exchange rates were impractical but, as a concession, he would announce that those interested in discussing floating rates could hold a separate session on the afternoon reserved for tennis, golf, swimming, and other recreation. (Meetings of this kind avoided shabby locations.) We declined.

This occurred after the Bretton Woods system had taken a major step toward its demise. In March 1968, President Johnson had embargoed gold exports except for central banks and discouraged central banks from asking for gold. Three years later, the dollar floated, temporarily officials believed. Nevertheless, it floated—and experience with floating showed that it was not impractical or destabilizing. The attempt to fix exchange at new parities lasted less than 15 months. In March 1973, the dollar floated and, with the exception of a few brief periods, the United States allowed it to float freely. The European Central Bank also allows the euro to float. Of course, many other countries peg or intervene. Since 1972, international reserves, mainly dollars, increased at an 8.5 percent compound annual rate. At the end of 2002, two-thirds of the $1.8 trillion in international reserves belonged to China, Japan, Taiwan, Korea, and Hong Kong.

Floating the dollar in 1971 was the most contentious issue when President Nixon met with his advisers at Camp David in August 1971. Arthur Burns opposed even after the president decided tentatively on including a floating rate in the package. The alternative was direct controls on imports, perhaps making the 15 percent surtax on imports, approved at that meeting, a permanent instead of temporary part of the president's program or imposing more so-called voluntary quotas on automobile, steel, and other imports.

Floating exchange rates increased freedom. Once the dollar floated permanently, the government removed capital controls that had been imposed in the 1960s. Other countries followed. Early in the 1980s, Britain held its first election without capital controls since 1936.

Ending the Military Draft. Milton Friedman was a proponent of an all-volunteer army and served as a member of the Gates Commission, which voted unanimously to recommend that President Nixon ask Congress to end the military draft. The idea was not new. The United States had relied on a volunteer army through much of its history. But relying on a volunteer army to fight a large-scale war was new, and there were many skeptics.

Walter Oi (1998) explained that President Nixon discussed a volunteer army during the 1968 election campaign. After the election Alan Wallis urged Arthur Burns to discuss the issue with President-elect Nixon. Wallis got Bill Mechling, Martin Bailey, Walter Oi and Harry Gilman to work out estimates of the demand and supply schedules and to compute equilibrium wages. Friedman's principal role was to explain and defend the proposal and respond to questions. As a member of the Gates Commission, appointed by President Nixon to consider alternatives to the military draft, he succeeded in getting the military and members of Congress to see the benefits to them and the nation of a volunteer army.

The war had become unpopular, and the draft had provoked riots, burning draft cards, and a spreading belief that the draft was unfair. An eligible draftee could escape by going to college, fleeing to Canada, or having braces put on his teeth. Ending the draft would be popular. The principal alternative to the voluntary army, however, was a lottery. An eligible citizen would get the privilege of serving his country if he lost—that is won—the lottery.

More than thirty years later, the volunteer army remains. A voice now and again calls for a return of the military draft—usually based on an effort to share the cost of war "equally," whatever that might mean. These voices do not, as far as I know, include senior military officers or presidents of either party.

Repealing Interest Rate Ceilings. Unlike the draft and floating exchange rates, there was little controversy among economists about repeal of interest rate ceilings. Those who wrote or spoke on that issue generally favored repeal of ceilings on time deposit rates. Within the Federal Reserve, there was considerable sentiment, possibly a majority in the 1950s, in favor of making the ceiling rate nonbinding (Meltzer 2003, Chapter 3). The members were reluctant to ask Congress to repeal the legislation. Although several recognized that the ceilings distorted allocation of financial assets, it was never exactly the right time to put ceiling rates on standby. Fear of congressional response appears to have been a major reason, and there was no sign of crisis. By the mid-1960s, members of the Federal Reserve acquiesced in or favored extension of ceiling rates to savings accounts at nonbank thrift institutions (Meltzer 2003, Chapter 4). The first response was to extend controls, not ease them. A first constructive step in 1970 removed interest rate ceilings from negotiable certificates of deposit (CDs) of $100,000 or more following the Penn Central failure.

It did not take long for entrepreneurs to recognize that fortunes could be made by organizing mutual funds to buy large CDs and, for a fee, offering participation to small depositors. The crisis came when the drain into these mutual funds became relatively large. Banks and thrifts had to buy back their deposits at interest rates well above the ceilings. This was particularly hard for mortgage lenders when short-term rates remained above the long-term rates on their mortgage portfolios. Elimination of ceilings had been widely discussed. Additional controls seemed unlikely to solve the problem. At best, they would postpone a permanent solution.

The three successes have in common that policymakers perceived that there was a crisis. There was also a well-presented, market alternative that dominated other available alternatives that offered less freedom to choose in markets.

There is some symmetry. There were no important crises affecting right to work laws, state universities, licensing of doctors and lawyers, and many other issues where the Friedmans proposed changes. A recent perceived crisis about accounting processes and accountants led to increased regulation, but there was no well-developed proposal to use markets and incentives in place of regulation. Political response is most often the work of lawyers. Their training typically leads them to propose regulations—command and control schemes—instead of aligning private and social interests by changing incentives when the two diverge.

NON-CRISIS CHANGES

Many of the Friedmans' proposals have been adopted in part without the push from a major crisis. Crises provide an opportunity for governments to make reforms, but reforms and changes occur at other times. Crises can create the sufficient condition for ideas and persuasion to influence policy, but the influence of ideas on policy occurs at other times.

Tariffs have been reduced throughout the world. Reductions are rarely unilateral, as the Friedmans urged, and trade agreements contain many protectionist clauses. Deregulation of transport, banking, and telecommunications is widely accepted as beneficial to consumers, but radio and television licensing remains regulated by the Federal Communications Commission. The Treasury offers inflation-indexed bonds, and it auctions long-term securities, as Friedman proposed in the 1960s. One result of the bond auction is that the Federal

Reserve ended its so-called even keel policy of supplying reserves to hold money market conditions constant when the Treasury marketed its debt. The even keel policy often required the Federal Reserve to supply enough bank reserves to ensure that the Treasury's issues were sold. During the 1960s and 1970s, reserves provided under even keel were a main source of excess money growth.

Education Vouchers. The Friedmans' proposal to issue education vouchers would increase parents' freedom to choose the school they believe benefits their children. Although many reports with titles such as "A Nation at Risk" use the language of crisis, the public, and its elected representatives, have given only modest support to Friedman-type vouchers. One reason is that until recently the constitutionality of parents using vouchers in parochial schools had not been decided. But vouchers and other methods that provide greater choice and increase pressure for reform of ineffective schools have not produced dramatic improvements in learning, as measured by conventional testing procedures. Further, proponents of vouchers would cite the intense opposition of teachers unions as part of the explanation for the failure of vouchers to be adopted widely.

I find the last argument incomplete. Interest groups did not prevent deregulation and increased choice in many areas. I do not question that teachers unions oppose vouchers. Why are they successful when others are not? Why did President Bush abandon a modest voucher proposal in his education bill?

I suggest that attitudes toward public goods and redistribution have an important role. Evidence suggests that some families purchase better schools by buying housing in districts known to have higher standards of educational achievement. They pay indirectly for school choice, so they have strong incentives to maintain their schools and much weaker incentives to improve schooling and learning for others. Also, proposals to give vouchers to the lowest income groups do not appeal to those just above the margin, especially if they pay full tuition to the parochial schools where many of the vouchers would be used.

Emphasis on increased choice fostered change. Both the charter school movement and home schooling have grown rapidly. In different ways, these two programs have enhanced opportunities for individual initiative and innovation in educational methods. The voucher proposal has encouraged choice and change in ways the Friedmans did not propose.

Negative Income Tax. One of the Friedmans' best-known proposals called for replacing all welfare programs with a negative income tax. Below a certain level of income, the tax authority would pay the citizen.

No government has replaced the welfare system with a negative income tax. The proposal continues to stimulate research, and it has been adopted as a supplement to existing welfare programs in various ways. The earned income tax credit (EITC) is one familiar program that supplements earned income for workers below median income.

There are three points to notice about EITC. First, it supplements programs like food stamps, health care, and educational grants but does not replace them. Second, it goes to people who have earned income and excludes those who do not. Third, the transfer payment increases with the number of dependent children, so it does not depend only on earnings. In 2000, 55 million people received payments under the EITC (Moffitt 2003). The maximum income at which a family received the transfer reached $32,000 in 2001, very close to median income of black families.

As Moffitt (2003, 32) notes in his recent survey, the negative income tax has ambiguous effects on labor supply and lacks a work requirement. Unlike transfers to farmers and business firms, most welfare programs either offer transfers in-kind or require work. Meltzer and Richard (1985) showed that a utility-maximizing voter prefers to offer in-kind transfers rather than cash transfers because the former induce more work by the recipients than do the latter. Public discussion of welfare-to-work programs during welfare reform in the 1990s, both in the United States and abroad, suggests that voters favor work requirements for all but the aged, infirm, and the most handicapped.

Again, on an issue involving substantial redistribution, the Friedmans' proposal influenced subsequent discussion but was not adopted in its entirety. The Meltzer and Richard paper developed a condition under which a negative income tax would replace in-kind transfers. The condition is that the pivotal or median voter does not work. Fortunately, we are not there yet.

I agree with Moffitt (2003, 33), who concluded his survey by noting that "the negative income tax has played a substantial role in reorienting the thinking of policymakers to the basic message that incentives matter. While this insight does not surprise academic economists, it is a new development in policy circles."

MONETARY POLICY

Possibly the most famous Friedman proposal called for a rule for constant monetary growth. Once again, Friedman's proposal had a major influence on subsequent developments but was not itself adopted. Several countries have an inflation target for monetary policy but not an explicit monetary rule. Absence of a consensus among economists is a major reason.

The history of money has a strong cyclical component. Periods in which money is considered a principal determinant of inflation are followed by the opposite—money is regarded as irrelevant for inflation or fluctuations in output. Currently, the most widely used model has only one interest rate: a short-term rate set by the central bank. Given that interest rate, aggregate demand and a Phillips curve determine output and inflation. With the interest rate fixed, the demand for money determines how much money the central bank supplies.

This model puts some strong restrictions on the relation of the interest rate fixed by the central bank to other asset prices and exchange rates in the short run, before full adjustment of asset and output markets. Friedman's (1956) analysis of the demand for money included these asset prices as separate arguments in the demand function. In his model, money can affect a wide spectrum of asset prices and components of output in the short run, and, in turn, spending and the demand for money can be affected by those relative prices and demands. Friedman's money growth rule avoided the as-yet intractable problem of predicting how the many relative prices or interest rates interact.

Discussion of monetary rules revived after the major inflation of the 1970s and the publication of Kydland and Prescott's (1977) paper on rules and discretion. That paper and subsequent work on credibility heightened and focused central bankers' concern for the public's anticipations and beliefs. A monetary rule or guide provided that information by increasing information about intended future policy.

Several central banks have now adopted a rule called inflation targeting. Many central banks calculate the interest rate consistent with their inflation and output targets using a formula or rule proposed by Taylor (1993), although they may not follow the rule. These limits on discretion are steps in the direction Friedman proposed.

Friedman's rule was a type of inflation control or inflation targeting. It differed from current inflation targeting rules not only by making money growth the key variable for central bank control but also by defining inflation as the *maintained* rate of price change. Friedman's much-quoted dictum that inflation is always and everywhere a monetary phenomenon excludes one-time price level changes. One should not doubt that Friedman understood that measured changes in a published price index could result from changes in tariffs, excise tax rates, exchange rates, oil shocks, productivity changes, and many other factors. His inflation rule permitted these price changes, positive and negative, to remain. The reported price level would be a random walk, but the expected rate of inflation would always be zero (or some constant value) if the maintained growth rates of output and monetary velocity remained unchanged. Rational individuals would make their consumption-saving decisions on this assumption. The Friedman rule would avoid persistent inflation and deflation and the wealth transfers they caused.

Current rules for inflation targeting treat all price level changes as inflation. There is no distinction between permanent, or persistent, changes and transitory, or one-time, price level changes. Following the constant inflation target requires rolling up or back all one-time price changes. In practice, many central banks will deviate from their target in the short run, if the target requires loss of output. The reason is that the difference between potential and actual output—the output gap—is an argument in economists' statement of the central banker's (or the community's) objective function.

HAS THE TIDE TURNED?

Writing in 1979–80 after the election of Margaret Thatcher, the Friedmans were optimistic about a retreat from the increased role for government that had characterized the previous fifty years in the United States and Great Britain. On the back cover of the 1990 edition, they were almost exultant. They ended *Free to Choose* this way (1980, 309–10):

> Fortunately, we are waking up. We are again recognizing the dangers of an over-governed society, coming to understand that good objectives can be perverted by bad means, that reliance on the freedom of people to control their own lives in accordance with their own values is the surest way to achieve the full potential of a great society.

Would that it were so! The picture is much more mixed in the quarter century since they first wrote. Tax rates have been reduced in the United States and many other countries. Tariffs are lower, permitting trade to expand and living standards to rise in many developing countries. Democratic choice of government has spread to places where it had never been known. The former Soviet Union collapsed, freeing many of its citizens and the citizens of its satellites to choose capitalism and democracy. Countries everywhere have adopted some of the Friedmans' proposals in whole or part. For example, Russia has privatized land and property, adopted a flat tax, and developed a pension system with less redistribution than ours.

Against these promising reasons for hope, there are ample reasons for concern. Most of the deregulation in the United States came before the Reagan administration. Government programs in education, health care, and retirement continue to expand. Even the current conservative administration promotes a massive expansion of government involvement in health care. Measuring government size by the ratio of government transfer payments to gross domestic product shows a rise from 4 percent during the Truman administration to approximately 13 percent early in 2003. Figure 1 shows that periods of largest growth occurred during the administrations of Presidents Eisenhower, Johnson, Nixon, and the two Bushes. The Nixon years are especially notable. The transfer ratio rose from about 6 percent to 9 percent. The administration did little to stop or slow growth of the Great Society programs and even added its own large program called revenue sharing. The few periods of relative decline in transfers came during the Truman, Kennedy, Reagan, and Clinton administrations. Relative decline can occur, of course, if GDP grows rapidly, as during the Clinton administration, but that administration also reduced welfare payments by reforming the welfare system.

Figure 1 does not include the proposed prescription drug program. Since it includes only the budgetary cost of transfers, it also excludes costs of in-

Figure 1
Transfer Payments as a Percent of Gross Domestic Product, 1946:1 to 2003:1
Percent

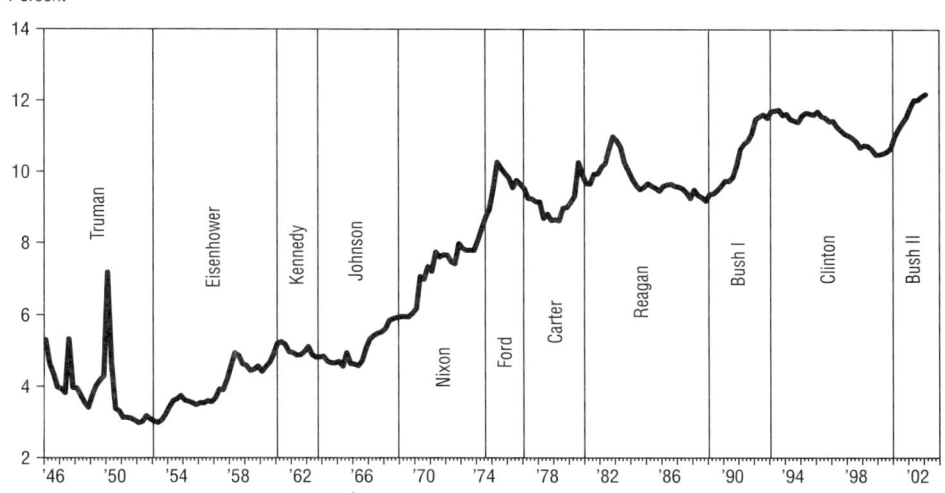

creased regulation. Government intrusiveness is harder to measure than transfers; I know of no quantification of the welfare cost of environmental, health, product safety, and financial regulation. Some examples are instructive. Environmental regulation of wetlands curtails property rights without paying compensation. Price-setting for medical services distorts resource use, taxes specific types of income, and transfers wealth arbitrarily. The list is long. Recently states' attorneys general have added to the list using the courts instead of the legislature to regulate and penalize.

Governments at all levels have reversed their position on racial discrimination. Instead of enforcing segregation, as in the past, the law now enforces desegregation. Freedom to choose surely increased as a result, but pressures for mandating equal outcomes, not just equal opportunity, increased also.

Many of the changes away from freedom to choose toward increased regulation and more transfers are prompted by appeal to fairness or equity. *Free to Choose* and *Capitalism and Freedom* have very little to say about fairness. Their usual explanation of continued and enhanced regulation and redistribution is either a version of "rational ignorance" or a powerful bureaucracy foisting its will, or its clients' will, on the public. The literature of political economy or public choice is mentioned but not exploited.

I accept that rational ignorance applies to many details of tax or regulatory legislation and interpretation. Hardly anyone, including the legislators, reads the often hundreds of pages in many of these bills. Few would understand the implications of many of the provisions for special interests, and much legislation allows the bureaucracy to fill in the details and decide on the application.

I submit that rational ignorance fails as an explanation of the trends in transfer payments or regulatory programs. Major spending programs for retirement, health, and education do not require knowledge of details to know whether one pays or receives. It does not take either much knowledge or much experience to know that governments cannot produce health care, pensions, or education. Production requires technical skills. Government provides inter- and intragenerational redistribution of the cost. When the government announces a major new program, such as the prescription drug program, those who benefit and those who pay generally know which group they are in. Entrepreneurial politicians understand that also. They recognize, for example, that the number of those who bear the cost of old age pensions is growing relative to those who benefit. It becomes feasible to discuss privatization for a small part of the program.

The first concern politicians typically have about a proposal is not whether it moves society toward a Pareto optimal allocation or increases efficiency. Their first concern is who gains and who loses, who pays and especially who receives. And that is what is most emphasized when they run for office. A Congressman who points to the new health benefit or new highway he got for the district doesn't point out that there are 434 other districts that received the same or possibly larger benefits.

By concentrating on the economic benefits and neglecting the political system and its interaction with the economic system, *Free to Choose* concludes on a more optimistic note than ex post judgment warrants. As de Tocqueville recognized long ago, the political incentive to redistribute income remains strong in democratic-capitalist systems. Votes are more equally distributed than income; each adult has one vote. Even if we allow for those who do not vote, the mass of the distribution of votes lies below the median earned income (Meltzer and Richard 1981). Voters with incomes at the median or below gain by transferring income to themselves. The efficiency loss to society from redistribution is part of the cost of democratic government and political freedom.

The Friedmans swam against this strong current. They could not stop or reverse it, but they influenced far more than most the ways in which people and politicians think and act. They influenced the economics profession and other scholars to analyze government programs and show their net costs. And they taught a generation or more about the value to them of remaining free to choose.

REFERENCES

Friedman, Milton. 1953. "The Case for Flexible Exchange Rates." In *Essays in Positive Economics*, ed. M. Friedman. Chicago: University of Chicago Press.

———. 1956. "The Quantity Theory of Money—A Restatement." In *Studies in the Quantity Theory of Money*, ed. M. Friedman. Chicago: University of Chicago Press, 3–21.

———. 1962. *Capitalism and Freedom*. Chicago: University of Chicago Press.

Friedman, Milton, and Rose D. Friedman. 1980. *Free to Choose: A Personal Statement*. New York: Harcourt Brace Jovanovich.

Hayek, F. A. 1944. *The Road to Serfdom*. Chicago: University of Chicago Press.

Kydland, Finn, and Edward C. Prescott. 1977. "Rules Rather Than Discretion: The Inconsistency of Optimal Plans." *Journal of Political Economy* 85 (June): 473–92.

Meltzer, Allan H. 2003. *A History of the Federal Reserve*, volume 2 (unpublished).

Meltzer, Allan H., and Scott F. Richard. 1981. "A Rational Theory of the Size of Government." *Journal of Political Economy* 89, 214–27. Reprinted as Chapter 2 in *Political Economy*, ed. A. Meltzer, A. Cukierman, and S. F. Richard. New York: Oxford University Press, 1991.

———. 1985. "A Positive Theory of In-Kind Transfers and the Negative Income Tax." *Public Choice* 47: 231–65. Reprinted as Chapter 4 in *Political Economy*, ed. A. Meltzer, A. Cukierman, and S. F. Richard. New York: Oxford University Press, 1991.

Moffitt, Robert A. 2003. "The Negative Income Tax and the Evolution of U.S. Welfare Policy." NBER Working Paper no. 9751 (unpublished).

Oi, Walter. 1998. "Historical Perspectives on the All-Volunteer Force: The Rochester Connection." In *Professionals on the Front Line*, ed. J. Eric Fredland et al., 37–54. Washington, DC: Brassey's.

Schumpeter, Joseph A. 1942. *Capitalism, Socialism, and Democracy*. New York: Harper & Brothers.

Schwartz, Anna J. 1993. In "Milton, Money, and Mischief: Symposium and Articles in Honor of Milton Friedman's 80th Birthday," ed. J. L. Jordan. *Economic Inquiry* 21 (April): 206–10.

Stigler, George J. 1975. "The Intellectual and His Society." In *Capitalism and Freedom: Problems and Prospects*, ed. R. T. Selden. Charlottesville: University Press of Virginia.

Taylor, John B. 1993. "Discretion Versus Policy Rules in Practice." *Carnegie Rochester Conference Series on Public Policy* 39, 195–214.

Friedman's Monetary Framework: Some Lessons

Ben S. Bernanke

It is an honor and a pleasure to have this opportunity, on the anniversary of Milton and Rose Friedman's popular classic, *Free to Choose*, to speak on Milton Friedman's monetary framework and his contributions to the theory and practice of monetary policy. About a year ago, I also had the honor, at a conference at the University of Chicago in honor of Milton's ninetieth birthday, to discuss the contribution of Friedman's classic 1963 work with Anna Schwartz, *A Monetary History of the United States*.[1] I mention this earlier talk not only to indicate that I am ready and willing to praise Friedman's contributions wherever and whenever anyone will give me a venue but also because of the critical influence of *A Monetary History* on both Friedman's own thought and on the views of a generation of monetary policymakers.

In their *Monetary History*, Friedman and Schwartz reviewed nearly a century of American monetary experience in painstaking detail, providing an historical analysis that demonstrated the importance of monetary forces in the economy far more convincingly than any purely theoretical or even econometric analysis could ever do. Friedman's close attention to the lessons of history for economic policy is an aspect of his approach to economics that I greatly admire. Milton has never been a big fan of government licensing of professionals, but maybe he would make an exception in the case of monetary policymakers. With an appropriately designed licensing examination, focused heavily on the fine details of the *Monetary History*, perhaps we could ensure that policymakers had at least some of the appreciation of the lessons of history that always informed Milton Friedman's views on monetary policy.

Today I will pass over Friedman's contributions to our knowledge of monetary history and focus instead on how his ideas have influenced our understanding both of how monetary policy works and how it should be used. That is, I will discuss both the *positive* and the *normative* implications of Friedman's thought. The usual disclaimers apply—that is, I speak for myself and not necessarily for my colleagues at the Federal Reserve.

In preparing this talk, I encountered the following problem. Friedman's monetary framework has been so influential that, in its broad outlines at least, it has nearly become identical with modern monetary theory and practice. I am reminded of the student first exposed to Shakespeare who complained to the professor: "I don't see what's so great about him. He was hardly original at all. All he did was string together a bunch of well-known quotations." The same issue arises when one assesses Friedman's contributions. His thinking has so permeated modern macroeconomics that the worst pitfall in reading him today is to fail to appreciate the originality and even revolutionary character of his ideas, in relation to the dominant views at the time that he formulated them.

To illustrate, I begin with the descriptive or positive side of Friedman's work on monetary policy. Here is a short summary of Friedman's own list of eleven key monetarist propositions, as put forth in the conclusion to his 1970 (note well that date) lecture, "The Counter-Revolution in Monetary Theory." These propositions are a reasonable description, I believe, of Friedman's basic views on how money affects the economy. Here they are (in my summary of slightly more detailed language in the original):

1. There is a consistent though not precise relationship between the rate of growth of money and the rate of growth of nominal income.
2. That relationship is not obvious, however, because there is a lag between money growth and nominal income growth, a lag that itself can be variable.
3. On average, however, the lag between money growth and nominal income growth is six to nine months.
4. The change in the rate of nominal income growth shows up first in output and hardly at all in prices.
5. However, with a further lag of six to nine months, the effects of money growth show up in prices.
6. Again, the empirical relationship is far from perfect.
7. Although money growth can affect output in the short run, in the long run output is determined strictly by real factors, such as enterprise and thrift.
8. Inflation is always a monetary phenomenon, in the sense that it can be produced only by money growth more rapid than output. However, there are many possible sources of money growth.
9. The inflationary impact of government spending depends on its financing.
10. Monetary expansion works by affecting prices of all assets, not just the short-term interest rate.

11. Monetary ease lowers interest rates in the short run but raises them in the long run.

Let me emphasize again that these propositions reflected Friedman's view as of some thirty-five years ago. At the time, they were far from being the conventional wisdom, as suggested by the term "Counter-Revolution" in the essay's title. What do we make of these propositions today?

First, the empirical description of the dynamic effects of money on the economy given in the first six propositions would be viewed by most policymakers and economists today as being, as the British would put it, "spot on." As a minor illustration of this point, in my own academic research I contributed to a large modern econometric literature that has used vector autoregression and other types of time series models to try to quantify how monetary policy affects the economy. The economic dynamics estimated by these methods correspond very closely to those outlined in Friedman's propositions.

These methods confirm that a monetary expansion (for example) leads with a lag of one to two quarters to an increase in nominal income. Perhaps more importantly, as Friedman emphasized, the responses of the quantity and price components of nominal income have distinctly different timing. In particular, as Friedman told us, a monetary expansion has its more immediate effects on real variables such as output, consumption, and investment, with the bulk of these effects occurring over two to three quarters. (I was going to say, as Friedman *first* told us, but perhaps the credit for that should go to David Hume. Friedman's work is, after all, part of a long and great tradition of classical monetary analysis.) These real effects tend to dissipate over time, however, so that at a horizon of twelve to eighteen months the effects of a monetary expansion or contraction are felt primarily on the rate of inflation. The same patterns have been found in empirical studies for virtually all countries, not only by vector autoregression analysis but by more structural methods as well. They are reflected in essentially all contemporary econometric models used for forecasting and policy analysis, such as the FRBUS model at the Federal Reserve. The lag between monetary policy changes and the inflation response is the reason that modern inflation-targeting central banks, such as the Bank of England, set a horizon of up to two years for achieving their inflation objectives.

Thus Friedman's description of the economic dynamics set in train by a monetary expansion or contraction, summarized in his first six propositions, has been largely validated by modern research. What about the other propositions? Friedman's seventh point, that money affects real outcomes in the short run but that in the long run output is determined entirely by real factors, such as enterprise and thrift, is of particular importance for both theory and policy. The proposition that money has no real effects in the long run, referred to as the principle of long-run neutrality, is universally accepted today by monetary economists. When Friedman wrote, however, the conventional view held that monetary policy could be used to affect real outcomes—for example, to lower the rate of unemployment—for an indefinite period. The idea that monetary policy had long-run effects—or, in technical language, that the Phillips curve relationship between inflation and unemployment could be exploited in the long run—proved not only wrong but quite harmful. Attempts to exploit the Phillips curve trade-off, which persisted despite Friedman's warnings in his 1968 presidential address to the American Economic Association, contributed significantly to the Great Inflation of the 1970s—after the Great Depression the second most serious monetary policy mistake of the twentieth century.

The diagnosis of inflation in Friedman's eighth proposition, also controversial when he wrote it, is likewise widely accepted today. Of course, as we all know, Friedman noted the close connection between inflation and money growth, though carefully acknowledging that excessive money growth could have many causes. As Milton and Rose discussed in Chapter 9 of the 1980 edition of *Free to Choose*, popular views in the 1960s and 1970s (and even the views of some Federal Reserve officials) held that inflation could arise from a variety of nonmonetary sources, including the power of unions and corporations and the greediness of oil-producing countries. An unfortunate implication of these views, whose deficiencies were revealed by bitter experience under President Nixon, was that wage–price controls and other administrative measures could successfully address inflation. We understand today that the Great Inflation would simply not have been possible without the excessively expansionist monetary policies of the late 1960s and 1970s.

Some of Friedman's descriptive propositions remain the subject of active research. For example, much research has investigated both theoretically and empirically the interactions of fiscal policy, monetary policy, and inflation. Friedman's view that fiscal deficits are inflationary only if they result in money creation, his ninth proposition, remains broadly accepted, but work by scholars such as Thomas Sargent, Neil Wallace, and Michael Woodford has shown that these links can be subtle. For example, Sargent and Wallace's "unpleasant monetarist arithmetic" suggested that a near-term tightening of monetary policy, by making the long-term fiscal situation less tenable, could (in principle at least) lead to inflation because the public will anticipate that the fiscal deficit must be financed eventually by money creation. More recently, Woodford's fiscal theory of the price level suggests that nonsustainable fiscal policies can drive inflation, even if the central bank resists monetization. Following Woodford, Olivier Blanchard has recently argued that tight money policies in Brazil, by raising the government's financing costs and thus worsening the fiscal situation, might have had inflationary consequences. Although this subsequent work has refined our understanding of the relationship between monetary and fiscal policy, these analyses are not inconsistent with the spirit of monetarist propositions, which place the blame for inflation on overissuance of nominal government liabilities.

Another area of pressing current interest derives from Friedman's tenth proposition, that monetary policy works by affecting all asset prices, not just the short-term interest rate. This classical monetarist view of the monetary transmission process has become highly relevant in Japan, for example, where the short-term interest rate has reached zero, forcing the Bank of Japan to use so-called quantitative easing methods. The idea behind quantitative easing is that increases in the money stock will raise asset prices and stimulate the economy, even after the point that the short-term nominal interest rate has reached zero. There is some evidence that quantitative easing has beneficial effects (including evidence drawn from the Great Depression by Chris Hanes and others), but the magnitude of these effects remains an open and hotly debated question.

The only aspect of Friedman's 1970 framework that does not fit entirely with the current conventional wisdom is the monetarists' use of money growth as the primary indicator or measure of the stance of monetary policy. Clearly, monetary policy works in the first instance by affecting the supply of bank reserves and the monetary base. However, in the financially complex world we live in, money growth rates can be substantially affected by a range of factors unrelated to monetary policy per se, including such things as mortgage refinancing activity (in the short run) and the pace of financial innovation (in the long run). Hence, it would not be safe to conclude (for example) that the recent decline in M2 is indicative of a tight-money policy by the Fed.

The imperfect reliability of money growth as an indicator of monetary policy is unfortunate because we don't really have anything satisfactory to replace it. As emphasized by Friedman (in his eleventh proposition) and by Allan Meltzer, nominal interest rates are not good indicators of the stance of policy, as a high nominal interest rate can indicate either monetary tightness or ease, depending on the state of inflation expectations. Indeed, confusing low nominal interest rates with monetary ease was the source of major problems in the 1930s, and it has perhaps been a problem in Japan in recent years as well. The real short-term interest rate, another candidate measure of policy stance, is also imperfect because it mixes monetary and real influences, such as the rate of productivity growth. In addition, the value of specific policy indicators can be affected by the nature of the operating regime employed by the central bank, as shown for example in empirical work of mine with Ilian Mihov.

The absence of a clear and straightforward measure of monetary ease or tightness is a major problem in practice. How can we know, for example, whether policy is "neutral" or excessively "activist"? I will return to this issue shortly.

Besides describing the effects of money on the economy, Friedman also made recommendations for monetary policy—the normative part of his framework. I will discuss just three of the most important of these.

First, Friedman has emphasized the Hippocratic principle for monetary policy: "First, do no harm." Chapter 9 of *Free to Choose* contains a famous quote of John Stuart Mill, as follows: "Like many other kinds of machinery, (money) only exerts a distinct and independent influence of its own when it gets out of order." On this quote, Milton and Rose commented: "Perfectly true, as a description of the role of money, provided we recognize that society possesses hardly any other contrivance that can do more damage when it gets out of order."

Friedman's emphasis on avoiding monetary disruptions arose, like many of his other ideas, from his study of U.S. monetary history. He had observed that, in many episodes, the actions of the monetary authorities, despite possibly good intentions, actively destabilized the economy. The leading case, of course, was the Great Depression, or as Friedman and Schwartz called it, the Great Contraction, in which the Fed's tightening in the late 1920s and (most importantly) its failure to prevent the bank failures of the early 1930s were a major cause of the massive decline in money, prices, and output. It is likely that Friedman's study of the Depression led him to look for means, such as his proposal for constant money growth, to ensure that the monetary machine did not get out of order. I hope, though of course I cannot be certain, that two decades of relative monetary stability have not led contemporary central bankers to forget the basic Hippocratic principle.

A second normative recommendation, worth recalling here, was Friedman's preference for floating rather than fixed exchange rates. At times, at least in popular writing, Friedman rationalized this position as following from free market principles. This argument is a bit disingenuous, I think, as a fixed nominal exchange rate is just one method of anchoring the aggregate price level and is perfectly consistent with free adjustment of the relative prices of goods and services. In a more serious vein, Friedman understood that, in a world in which monetary policymakers put domestic economic stability above balance of payments considerations, a fixed exchange rate system is likely to be unstable during periods of economic stress. He saw that this was the case during the 1930s, when the world was on a modified gold standard called the gold exchange standard, and it was likewise the case under the postwar Bretton Woods system. To reconcile a fixed exchange rate and an emphasis on domestic stability, policymakers must impose capital controls or restrictions on trade, which have undesirable effects on economic efficiency.

If policymakers' first priority is stability of the domestic economy, Friedman reasoned, then why not adopt a system—namely, flexible exchange rates—that provides the necessary monetary independence without restrictions on the flow of capital or goods? When Friedman wrote about fixed and flexible exchange rates, a switch from the Bretton Woods fixed-exchange-rate system to a floating-rate system seemed quite unlikely. In this as in many other matters, he was prescient, as the major currencies have now been successfully floating since the breakup of the Bretton Woods system in the early 1970s.

These two recommendations have had major effects on institutional design and policy practice. However, in my view, the most fundamental policy recommendation put forth by Milton Friedman is the injunction to policymakers to provide a stable monetary background for the economy. I take this to be a stronger statement than the Hippocratic injunction to avoid major disasters; rather, there is a positive argument here that monetary stability actively promotes efficiency and growth. (Hence Friedman's suggestion that the long-run Phillips curve, rather than vertical, might be positively sloped.) Also implicit in Friedman's focus on nominal stability is the view that central banks should avoid excessively ambitious attempts to manage the real economy, which in practice may exacerbate both nominal and real volatility. In Friedman's classic 1960 work, *A Program for Monetary Stability*, he suggested that monetary stability might be attained by literally keeping money stable: that is, by fixing the rate of growth of a specific monetary aggregate and forswearing the use of monetary policy to "fine-tune" the economy.

Do contemporary monetary policymakers provide the nominal stability recommended by Friedman? The answer to this question is not entirely straightforward. As I discussed earlier, for reasons of financial innovation and institutional change, the rate of money growth does not seem to be an adequate measure of the stance of monetary policy, and hence a stable monetary background for the economy cannot necessarily be identified with stable money growth. Nor are there other instruments of monetary policy whose behavior can be used unambiguously to judge this issue, as I have already noted. In particular, the fact that the Federal Reserve and other central banks actively manipulate their instrument interest rates is not necessarily inconsistent with their providing a stable monetary background, as that manipulation might be necessary to offset shocks that would otherwise endanger nominal stability.

Ultimately, it appears, one can check to see if an economy has a stable monetary background only by looking at macroeconomic indicators such as nominal GDP growth and inflation. On this criterion it appears that modern central bankers have taken Milton Friedman's advice to heart. Over the past two decades, inflation has fallen sharply and stabilized around the world, not only in the industrialized nations but in emerging-market economies and in even the poorest developing nations. Some central banks, so-called inflation targeters, have set explicit, quantitative targets for inflation; but all central banks, certainly including the Federal Reserve, have emphasized the importance of achieving and maintaining price stability. On the issue of inflation control, Friedman may be judged to have been a bit too pessimistic; his concerns that central banks would have neither the technical ability nor the correct incentives to control inflation led him to recommend his money-growth rule, for which a central bank could certainly be held accountable. Evidently, however, determined central banks can stabilize inflation directly; at least they have been able to do so thus far.

However, on the benefits of monetary stability, or as I would prefer to say, nominal stability, Friedman was not wrong. Many theories popular even today might lead one to conclude that increased stability in inflation could be purchased only at the cost of reduced stability in output and employment. In fact, over the past two decades, increased inflation stability has been associated with marked increases in the stability of output and employment as well, both in the United States and elsewhere.

It has been argued that a lower incidence of exogenous shocks explains these favorable developments, and that may be part of the story. But I believe that there is an important causal relationship as well. For example, low and stable inflation has not only promoted growth and productivity, but it has also reduced the sensitivity of the economy to shocks. One important mechanism has been the anchoring of inflation expectations. When the public is confident that the central bank will maintain low and stable inflation, shocks such as sharp increases in oil prices or large exchange rate movements tend to have at most transitory price-level effects and do not result in sustained inflationary surges. In contrast, when inflation expectations are poorly anchored, as was the case in the 1970s, shocks of these types can destabilize inflation expectations, increasing the inflationary impact and leading to greater volatility in both inflation and output.

In summary, one can hardly overstate the influence of Friedman's monetary framework on contemporary monetary theory and practice. He identified the key empirical facts and he provided us with broad policy recommendations, notably the emphasis on nominal stability, that have served us well. For these contributions, both policymakers and the public owe Milton Friedman an enormous debt.

NOTE

[1] See "On Milton Friedman's Ninetieth Birthday," by Ben S. Bernanke, University of Chicago, November 8, 2002, www.federalreserve.gov/boarddocs/speeches/2002/20021108/default.htm.

Session 6

What Have We Learned from the Measurement of Economic Freedom?
James Gwartney and Robert Lawson

Can the Tide Turn?
Raghuram G. Rajan

What Have We Learned from the Measurement of Economic Freedom?

James Gwartney and Robert Lawson

> *When you can measure what you are speaking about, and express it in numbers, you know something about it; but when you cannot measure it, when you cannot express it in numbers, your knowledge is of a meagre and unsatisfactory kind.*
> —Lord Kelvin, 1883

Milton Friedman is the godfather of the Economic Freedom of the World (EFW) project. Michael Walker of the Fraser Institute traces the origins of the project to the 1984 session of the annual meeting of the Mont Pelerin Society. In responding to the paper presented by the historian Paul Johnson, Walker quoted the following passage from one of Friedman's classic works, *Capitalism and Freedom*:

> [H]istorical evidence speaks with a single voice on the relation between political freedom and a free market. I know of no example in time or place of a society that has been marked by a large measure of political freedom, and that has not also used something comparable to a free market to organize the bulk of economic activity.

This quotation led to a rather heated discussion about the distinction between political and economic freedom. Over lunch following the session, Mike Walker convinced Milton and Rose Friedman to cohost a conference on this topic and soon thereafter persuaded Neil McLeod of the Liberty Fund of Indianapolis, Indiana, to provide the necessary funding. Walker recognized that Milton's participation in the project would make it possible to recruit a number of the world's finest minds to participate. He was right. Lord Peter Bauer, Gary Becker, Douglass C. North, Armen Alchain, Arnold Harberger, Alvin Rabushka,

Gordon Tullock, and Sir Alan Walters were among the early participants. The initial meeting led to a series of conferences between 1988 and 1994 that focused on the development of a clear definition of economic freedom and presentations concerning how it might be measured. The Economic Freedom of the World measure is a direct result of these conferences.

While I was not among those participating in the first two conferences, I read the proceedings of those sessions and began attending with the third of what eventually became a series of six meetings.[1] Participants looked to Milton for direction, and he certainly provided it. Two things made a vivid impression on me. First, Milton was convinced that despite the complex and multidimensional nature of economic freedom, it could be measured. Moreover, it was important to do so. He told conference participants that social scientists at the University of Chicago often argued, "If you can't measure it, measure it anyway." This view reflects the 1883 quotation from Lord Kelvin that opens this paper. Milton's position provided inspiration that a reasonably good measure of economic freedom could be developed.

Second, both Milton and Rose were constantly reminding us that our goal was the development of a scientific instrument that could be used to quantify the degree of economic freedom across a large number of countries. To the fullest extent possible, the measure needed to be based on objective data. We did not want our subjective views to influence the rating of any country. We wanted to develop an index that others could replicate and that even those who disagreed with us would utilize as a research tool.

CHARACTERISTICS OF THE EFW INDEX

The foundation of the EFW index is the proposition that individuals own themselves; they are not owned by the government. Because of this self-ownership, the protection of individuals and their property against aggression by others is the core of economic freedom. Of course, ownership also implies the right to enter markets and exchange goods and services with others at mutually agreeable terms. Thus, the four cornerstones of economic freedom are (1) private ownership, (2) personal choice, (3) voluntary exchange, and (4) free entry into markets. The EFW index is designed to measure the degree to which a nation's institutions and policies are consistent with these four cornerstones.

Initially, the EFW index was based on seventeen quantifiable components like government expenditures as a share of GDP (see Gwartney, Lawson, and Block 1996). This focus on objectively quantifiable variables led to a problem: Important legal and regulatory elements that influence economic freedom were omitted from the index. To correct this deficiency, additional components based on survey data were added to the index during 1997–2000.[2] As seen in the appendix to this article, the index now contains thirty-eight components (and subcomponents) that are divided into five major areas: (1) size of government, (2) legal structure and security of property rights, (3) access to sound money, (4) freedom to exchange with foreigners, and (5) regulation of credit, labor, and business. Country ratings for each of the thirty-eight components are derived on a zero-to-ten scale and then used to derive the summary and area ratings for the 123 countries covered by the index.[3] The EFW data are available for approximately 100 countries continuously (at five-year intervals) throughout 1980–2001.[4] Ratings are available for a smaller set of countries as far back as 1970. For additional details on how the components are transformed to a zero-to-ten scale and used to derive the area and summary ratings, see Gwartney and Lawson (2003) or the web site www.freetheworld.com.

For a country to achieve a high EFW rating, its government must do some things while refraining from others. Perhaps most important, the country's legal institutions must protect the property rights of owners and provide for the even-handed enforcement of contracts. Citizens must also be provided with access to sound money. This may be done by either following monetary policies that keep inflation at a low and steady rate or by removing obstacles that limit the use of alternative currencies. Like poorly defined property rights, polluted money retards voluntary exchange and thereby the exercise of economic freedom. But governments must refrain from other activities. When government spending, taxes, and regulations restrict exchange, limit entry into markets, and substitute regulations and mandates for private contracts, governments are limiting the economic freedom of their citizens.

In their best-selling book *Free to Choose*, the legacy of which is the focus of this conference, Milton and Rose Friedman proposed a number of amendments to the U.S. Constitution designed to preserve the economic liberty of Americans. These amendments included (1) a tax-spending limitation, (2) prohibition of duties on both imports and exports, (3) no price controls on either prices or wages, (4) no government licenses restricting entry into occupations and businesses, (5) the requirement that all direct taxes be levied at a flat rate, and (6) a money supply growth rule and inflation protection amendment designed to ensure that citizens have access to sound money. While the EFW index understandably provides more detail, it clearly reflects the same concept of economic freedom as that outlined by the Friedmans in their proposed economic bill of rights.[5]

WHAT HAVE WE LEARNED FROM THE EFW PROJECT?

Since the time of Adam Smith, market economy supporters have argued that countries that rely more heavily on markets to organize economic activity will grow more rapidly and achieve higher income levels than their more politically driven counterparts. Is this proposition really true? Without a comprehensive measure of economic freedom, the answer to this question is problematic.

When analyzing issues of growth and income levels, it is important to focus on a lengthy period of time. Short-term fluctuations in growth rates will be influenced by a number of factors, such as business cycle conditions and changes in the world price of important import or export items. In the short run, these largely random factors may dominate and conceal the strength of the relationship between economic freedom and growth. Moreover, credibility will influence the response to a policy change. Before decisionmakers will be willing to make major behavioral changes, they must be convinced that the change in policy direction is permanent rather than temporary. Furthermore, it will take time for information to be transmitted and markets to adjust fully to a new economic environment. Thus, the primary response to a policy change will often be delayed, and the full response will almost always be greater in the long run than in the short run.

Because the EFW data are both comprehensive and available over a lengthy time period, they are particularly suitable for the systematic analysis of cross-country differences in income levels and the long-run growth process. Using a database of the ninety-nine countries for which the EFW ratings were continuously available at five-year intervals during 1980–2000, the following tables summarize seven of the most important findings of this research.[6] The appendix at the end of this article contains variable definitions and sources as well as descriptive statistics for help in interpreting the table data.

1. The maintenance over a lengthy period of time of institutions and policies consistent with economic freedom is a major determinant of current cross-country differences in per capita GDP.

The mean EFW rating over the two decades of 1980–2000 provides a measure of long-term institutional quality. To achieve a high mean rating, a country would have to follow policies largely consistent with economic freedom throughout the lengthy period. Similarly, a low mean rating would be indicative of long-term policies inconsistent with economic freedom. As Table 1 shows (Equation 1), cross-country differences in the mean EFW rating during 1980–2000 explain 63.2 percent of the cross-country variation in 2000 per capita GDP.[7] When the percentage of a country's population residing in the tropics (a variable popularized by Jeffrey Sachs) is added to the model (Equation 2), the explanatory power increases to 75.1 percent. Clearly, long-term differences in institutional quality exert an enormous impact on per capita income levels.

2. An institutional and policy environment consistent with economic freedom is a key determinant of investment.

Table 2 shows the relationship between long-term economic freedom and various measures of investment after the effects of location, geography, and initial income level are taken into account. As Equation 1 shows, a one-unit increase in the mean EFW rating during 1980–2000 was associated with a $1,281 (1995 U.S. dollars) increase in annual real investment per worker during the two decades.

The investment per worker figures of Equation 1 include both private sector and public sector investment. Foreign direct investment (FDI) per worker provides an alternative measure that will be almost entirely reflective of private investment flows. Furthermore, the FDI figures will reflect the attractiveness of a country's investment climate to those residing outside of the country. As Equation 2 (Table 2) shows, economic freedom exerts a positive and significant impact on the inflow of foreign investment. A one-unit increase in EFW was associated with a $546 increase in annual FDI per worker during 1980–2000. A higher initial income level was associated with more foreign investment per worker, but the geographical variables did not exert a significant impact on FDI.

Equation 3 (Table 2) illustrates the impact of EFW on investment as a share of GDP (I/GDP). Once again, the EFW rating is positive and statistically significant. Other things constant, a one-unit increase in long-term EFW enhances investment as a share of GDP by 2.16 percentage points. Equation 4 (Table 2) considers the impact of economic freedom on the growth of capital per worker. Again, the impact is positive and significant. A one-unit increase in a country's mean EFW rating during 1980–2000 enhanced the annual growth rate of physical capital per worker by an estimated 1.24 percentage points.

Taken as a group, the regressions of Table 2 indicate that a country's insti-

Table 1
Economic Freedom and Cross-Country Differences in GDP per Capita

Dependent Variable: GDP per Capita, 2000
(t-ratio in parentheses)

Independent Variables	(1)	(2)
EFW rating, 1980–2000	651 (13.00)	529 (11.91)
Tropics		−8,472 (7.03)
Intercept	−11,183	−2,575
R^2 (adjusted)	63.2	75.1
Number of countries	99	99

tutional environment exerts a strong impact on capital formation. Countries that adopt policies and institutions that are consistent with economic freedom will have higher rates of capital formation, while countries that adopt unsound institutions will find that capital will flow elsewhere.

3. Economic freedom not only exerts an impact on the level of investment, it also influences growth by improving the productivity of investment.

Table 3 illustrates this point. The dependent variable in Table 3 is the growth of per capita GDP during 1980–2000. As Equation 1 shows, investment as a share of GDP exerts a highly significant positive impact on long-term growth. Equation 2 (Table 3) interacts the investment variable with economic freedom. The first independent variable multiplies I/GDP by one if a nation's

Table 2
Economic Freedom, Geography, and Location as Determinants of Investment

	Investment per Worker (US$), 1980–2000	FDI per Worker (US$), 1980–2000	I/GDP, 1980–2000	Growth of Kpw, 1980–1999
Independent Variables	(1)	(2)	(3)	(4)
EFW rating, 1980–2000	1,281 (4.12)	546 (4.00)	2.16 (3.09)	1.24 (3.76)
GDP per capita, 1980 (in 1000s US$)	834 (8.46)	122 (2.92)	−0.60 (2.70)	−0.51 (4.76)
Tropics	−563 (0.92)	−16 (0.06)	−3.76 (2.74)	−2.36 (3.69)
Coastal	−535 (0.83)	−43 (0.16)	3.00 (2.06)	0.53 (0.77)
Intercept	−6,457	−2,883	12.28	−2.04
R^2 (adjusted)	79.2	51.2	18.5	22.0
Number of countries	99	97	99	91

Note: Hong Kong and Taiwan were omitted from Equation 2 above because the FDI data were unavailable.

EFW rating is 7 or above and zero otherwise. The second independent variable does the same for nations with an EFW rating between 5 and 6.99, and the third independent variable separates out countries with an EFW rating below 5. The key feature of the regression is the relative size of the coefficients. For countries with EFW ratings of 7 or above, the coefficient is 0.275, which is greater than the 0.236 coefficient for the countries with ratings from 5 to 6.99, which in turn is greater than the 0.197 coefficient on the countries with ratings below 5. This indicates that a unit increase in investment enhances growth by a larger amount in countries with higher EFW ratings. This simple equation including only the interaction of EFW and investment explains almost half of the variation in GDP growth across countries.

Equation 3 (Table 3) adds the geographical variables to the model. The magnitudes of the coefficients for the I/GDP variables all fall by a small amount,

Table 3
Economic Freedom and the Productivity of Investment

Dependent Variable: Average Annual Growth Rate of GDP per Capita, 1980–2000
(t-ratio in parentheses)

Independent Variables	(1)	(2)	(3)
I/GDP, 1980–2000	0.244 (8.74)		
I/GDP, 1980–2000 x EFW > 7.0		0.275 (9.40)	0.242 (7.81)
I/GDP, 1980–2000 x 5.0 < EFW < 7.0		0.236 (8.76)	0.212 (7.56)
I/GDP, 1980–2000 x EFW < 5.0		0.197 (6.52)	0.183 (6.21)
Tropics			−0.937 (2.93)
Coastal			0.344 (0.83)
Intercept	−3.96	−3.72	−2.91
R^2 (adjusted)	43.5	49.7	53.1
Number of countries	99	99	99

but the pattern remains the same. The coefficient of 0.242 on the group with EFW ratings of 7 or more is 13.6 percent higher than the coefficient of 0.212 on the middle group of countries. Similarly, investment in the highest-rated group of countries is 31.7 percent more productive than in the lowest-rated group (EFW < 5.0), where productivity is measured as the impact that a given level of investment has on the rate of per capita GDP growth. Thus, investment is more productive—it exerts a stronger impact on growth—when it is undertaken in countries with higher EFW ratings.

4. When both direct (through improved efficiency) and indirect (through enhancement of the investment rate) effects are taken into account, a one-unit increase in EFW increases long-term growth by about one and a quarter percentage points.

Table 4 incorporates the key institutional, geographic-locational, and capital formation variables into combined models and uses them to analyze the growth of per capita GDP during 1980–2000. It also incorporates a methodology capable of capturing both the direct (through improvements in efficiency and productivity) and indirect (through capital formation) effects of economic freedom on the long-term growth of per capita GDP.

The dependent variable in Table 4 is the growth rate of real GDP per capita during 1980–2000. In Equation 1 the independent variables are mean EFW (1980–2000), changes in physical and human capital (Kpw and Hpw), percentage of population residing in the tropics (Tropics), percentage of population living within 100 kilometers of an ocean coastline (Coastal), and initial per capita GDP.[8] The independent variables all have the expected sign, and together they explain almost 60 percent of the cross-country variation in growth rates during 1980–2000.

The 0.81 coefficient on the EFW variable indicates that a one-unit difference in the 1980–2000 mean EFW rating is associated with 0.81 of a percentage point increase in growth during the period after the effects of the other independent variables, including Kpw, have been taken into account. Thus, the EFW coefficient of Equation 1 reflects only its direct impact on growth as a result of its impact on the efficiency of resource use. But this is only part of its impact on growth. As illustrated in Table 2, EFW also influences investment and the growth of the capital stock (Kpw). As Equation 4 of Table 2 shows, a one-unit increase in a country's EFW rating is associated with a 1.24 percent increase in the growth rate of a country's capital stock per worker. The EFW coefficient in Equation 1 of Table 4 will not reflect this indirect impact.

To capture both the direct (resulting from a more efficient use of resources) and indirect (resulting from higher investment levels) effects of EFW, the residuals from Equation 4 of Table 2 were substituted for the change in Kpw variable in the model of Table 4. These residuals indicate the cross-country variation in Kpw that is unrelated to EFW and the other independent variables of Equation 1 of Table 4. When this substitution is made, the coefficients for EFW and the other independent variables in the model will register both their direct and indirect (through changes in the growth of Kpw) effects on the growth of per capita GDP. Equation 2 (Table 4) indicates that the combined direct and indirect effects of a one-unit change in EFW enhance long-term growth by an estimated 1.24 percentage points.

Thus, higher institutional quality, as measured by the EFW rating, has two reinforcing effects on the relationship between investment and GDP growth:

Table 4
Economic Freedom, Investment, Geography, and Location as Determinants of Economic Growth

Dependent Variable: Average Annual Growth Rate of GDP per Capita, 1980–2000
(t-ratio in parentheses)

Independent Variables	(1)	(2)
EFW rating, 1980–2000	0.81 (4.00)	1.24 (6.67)
Growth of Kpw, 1980–1999	0.35 (5.70)	
Growth of Kpw, 1980–1999 (residuals)		0.35 (5.70)
Growth of Hpw, 1980–1999	0.42 (2.08)	0.42 (2.08)
Tropics	−1.30 (3.37)	−2.12 (5.90)
Coastal	0.49 (1.25)	0.68 (1.73)
GDP per capita, 1980 (in 1000s US$)	−0.16 (2.33)	−0.33 (5.58)
Intercept	−3.51	−4.21
R^2 (adjusted)	59.1	59.1
Number of countries	91	91

Note: The residuals for Growth of Kpw in Equation 2 above are from Table 2, Equation 4.

Better institutions both increase the level of investment and enhance its productivity. When both of these effects are taken into account, a one-unit change in EFW increases long-term growth by an estimated 1.24 percentage points. Because this is a change in a growth rate, it will have a large cumulative effect. Over a thirty-year period, for example, a one-unit increase in a country's EFW index would increase the country's per capita GDP by approximately 43 percent.

These findings illustrate why so much of the growth literature based on the production function approach of Solow is highly misleading. Until recently, almost all of the production function growth models failed to include institutional measures. Thus, they omitted both the direct and indirect effects of institutional quality. Moreover, even the more recent growth models that sometimes include various indicators of institutional quality along with investment fail to register the indirect effects of institutions. Thus, they continue to understate the importance of institutional quality (and economic freedom). Hopefully, incorporation of the EFW measure into the growth models of the future will help alleviate some of the misleading impressions created by the omissions of the past.

5. Changes in economic freedom enhance long-term growth.

Even though countries with higher EFW ratings grow more rapidly, some might still question whether changes in economic freedom enhance long-term growth. Table 5 sheds light on this issue. The dependent variable in Table 5 is the annual growth rate of per capita GDP during 1980–2000. In addition to the mean 1980–2000 EFW rating, changes in EFW during the decades of the 1980s and the 1990s are included in the analysis. Equation 1 includes the change in physical and human capital per worker (Kpw and Hpw), along with the three economic freedom variables. These five variables explain 58.5 percent of the cross-country variation in annual growth rates during 1980–2000.

The change in EFW during the 1980s exerted a positive and significant impact on the annual growth rate over the two decades. A one-unit increase in EFW during the 1980s enhanced growth during 1980–2000 by an estimated 0.71 percentage point. The change in EFW during the 1990s was positive but insignificant. The insignificance of the change during the 1990s is not surprising given the expected time lag accompanying an institutional change and the fact that a change during the 1990s would potentially impact growth for only a fraction of the two decades.

Equations 2 and 3 (Table 5) add two additional variables, tropical location and initial income level, that prior analysis suggests exert a significant impact on the growth of per capita GDP. The addition of these two variables increases the explanatory power of the model to 62.4 percent. Both the tropical and initial income variables are significant and have the expected sign, but they exert little impact on either the pattern or the significance of the other variables in the model. The change in EFW during the 1980s is significant in both Equations 2 and 3, and its estimated impact on the growth rate of per capita GDP remains near seven-tenths of a percentage point. The change in EFW during the 1990s continues to be positive, but it falls just short of significance at the 90 percent confidence level.

The pattern of these results sheds light on the impact of institutional change. The size and robustness of the change in EFW during the 1980s suggest that changes in institutional factors make a difference and that they will continue to exert an impact on economic growth over a long period of time. Correspondingly, the size and insignificance of the change in EFW during the 1990s indicate that the full impact of an institutional change will take time and that the immediate effects may be relatively small.

Table 5
Changes in Economic Freedom and Economic Growth

Dependent Variable: Average Annual Growth Rate of GDP per Capita, 1980–2000 (t-ratio in parentheses)

Independent Variables	(1)	(2)	(3)
EFW rating, 1980–2000	0.59 (4.17)	0.50 (3.38)	0.89 (4.35)
Change in EFW rating, 1980–1990	0.71 (3.09)	0.65 (2.84)	0.68 (3.08)
Change in EFW rating, 1990–2000	0.23 (1.34)	0.19 (1.13)	0.27 (1.62)
Growth of Kpw, 1980–1999	0.42 (7.67)	0.41 (7.54)	0.33 (5.69)
Growth of Hpw, 1980–1999	0.47 (2.33)	0.45 (2.23)	0.49 (2.51)
Tropics		−0.57 (1.86)	−1.15 (3.12)
GDP per capita, 1980 (in 1000s US$)			−0.17 (2.66)
Intercept	−4.15	−3.19	−4.40
R^2 (adjusted)	58.5	59.7	62.4
Number of countries	91	91	91

6. The EFW measure explains the divergence/convergence puzzle.

Economic theory suggests two major reasons why the income levels of less developed economies will converge toward their higher income counterparts. First, in a world of diminishing returns, the neoclassical model implies that the productivity of capital will be lower in high-income countries where capital is more plentiful than in lower income countries where it is more scarce. In turn, the higher productivity of capital in the low-income countries will cause capital to flow in their direction, thereby enhancing their growth and promoting the convergence of cross-country income levels. Second, low-income countries will be able to emulate and adopt, either freely or at a low cost, the proven technologies and successful business techniques employed in the more advanced nations. In contrast, new technologies and better business practices will have to be discovered, perhaps through costly research and development, in the more developed economies. Thus, technology and entrepreneurial activity should exert a more positive impact on growth in the less developed areas. This too should lead to convergence.

Table 6
Divergence, Convergence, and Economic Freedom

Dependent Variable: Average Annual Growth Rate of GDP per Capita, 1980–2000
(t-ratio in parentheses)

Independent Variables	**(1)**	**(2)**	**(3)**
GDP per capita, 1980 (in 1000s US$)	0.12 (2.25)	–0.14 (2.10)	–0.16 (2.30)
EFW rating, 1980–2000		1.28 (5.49)	0.87 (4.49)
Growth of Kpw, 1980–1999			0.35 (5.78)
Growth of Hpw, 1980–1999			0.44 (2.17)
Tropics			–1.24 (3.23)
Intercept	0.81	–5.38	–3.71
R^2 (adjusted)	4.0	26.2	58.8
Number of countries	99	99	91

Thus, traditional economic theory indicates that capital should move toward low-income countries and that these countries should grow more rapidly than their higher income counterparts. But the real world is inconsistent with this view; most low-income countries have grown less rapidly than their high-income counterparts. Income divergence rather than convergence is the norm (Dollar 1992, Pritchett 1997). Many economists have been puzzled by this phenomenon.

As Equation 1 of Table 6 shows, there was a positive and significant relationship between initial (1980) income level and the growth of per capita GDP during 1980–2000 for the ninety-nine countries of our basic data set. This positive relationship indicates that there was a tendency for high-income countries to grow more rapidly than those with low initial income levels. These findings reflect the divergence trend documented by others. However, Equation 2 (Table 6) illustrates the source of the divergence. Once the EFW variable is introduced into the model, the sign of the initial (1980) per capita GDP variable switches from positive to negative and the t-ratio (2.10) indicates significance at more than the 95 percent level of confidence. The expected trend toward income convergence is indeed present for countries with similar amounts of economic freedom. Further, as Equation 3 (Table 6) shows, this trend toward convergence is unaffected by the introduction of the physical capital, human capital, and tropical location variables into the model. Thus, when the consistency of a nation's institutions and policies with economic freedom is held constant, lower income countries grow faster than higher income countries, providing empirical support for models that predict convergence.[9]

7. A sound legal system is essential for sustained growth and achievement of high-income levels.

Protection of privately owned property and the evenhanded enforcement of contracts are essential ingredients for the achievement of prosperity. Without the legal protection of private property, the incentive of individuals to develop productive resources and engage in entrepreneurial activities is eroded. Correspondingly, without the enforcement of contracts, trade and the accompanying realization of gains from division of labor and specialization are stifled.[10]

The works of Douglass C. North and Friedrich von Hayek explain why a country's legal system is a vitally important determinant of its prosperity. Our modern living standards are the result of what North calls "depersonalized exchange," that is, trade between parties that do not know each other and will probably never meet. These exchanges are coordinated by what Hayek refers to as the "extension of the market" from the local town or village to the region, nation, and indeed to the far corners of the world. Almost everything that households in North America, Europe, and other parts of the developed world consume is the result of gains from depersonalized exchange and extension of the market. Without these gains, high levels of per capita income and modern

Table 7
Quality of Legal System, per Capita GDP, and Economic Growth

Countries with Average Legal Rating > 7.0 during 1980–2000	Legal System Rating	Per Capita GDP 2000 (US$)	Growth of per Capita GDP, 1980–2000
Switzerland	8.65	27,780	0.82
United States	8.61	33,960	2.12
Netherlands	8.58	26,910	1.98
New Zealand	8.51	17,840	1.29
Austria	8.49	26,420	1.99
Luxembourg	8.45	53,410	4.26
Denmark	8.41	28,680	1.74
Finland	8.36	24,160	2.27
Germany	8.36	25,100	1.70
Canada	8.32	26,840	1.69
Norway	8.31	29,200	2.42
Australia	8.29	24,550	1.96
Iceland	8.08	28,910	1.67
Sweden	8.05	23,650	1.66
Belgium	7.97	25,220	1.91
United Kingdom	7.91	23,580	2.29
Ireland	7.91	30,380	4.91
Singapore	7.89	23,700	4.92
Japan	7.84	25,280	2.34
Portugal	7.50	17,710	2.91
France	7.48	23,490	1.72
Hungary	7.16	11,960	1.31
Hong Kong	7.16	25,180	4.07
Taiwan	7.03	13,279	6.00
Average	**8.05**	**25,716**	**2.50**

Countries with Average Legal Rating < 4.0 during 1980–2000	Legal System Rating	Per Capita GDP 2000 (US$)	Growth of per Capita GDP, 1980–2000
Indonesia	3.90	2,970	3.69
Senegal	3.84	1,450	0.57
Sri Lanka	3.67	3,400	3.49
Pakistan	3.66	1,870	2.46
Honduras	3.62	2,830	−0.13
Syria	3.56	3,280	0.64
Iran	3.55	5,720	1.09
Nicaragua	3.54	2,450	−2.26
Peru	3.52	4,630	−0.24
Philippines	3.49	3,790	−0.02
Algeria	3.47	6,150	−0.20
Colombia	3.43	7,010	1.04
Uganda	3.42	1,450	2.23
Nigeria	3.34	860	−0.93
El Salvador	3.27	5,240	0.57
Congo, Republic of	3.27	950	0.37
Bolivia	3.20	2,310	−0.28
Bangladesh	3.19	1,540	2.57
Guatemala	3.02	4,430	−0.08
Haiti	2.98	1,920	−2.39
Congo, Democratic Republic of	2.38	730	−5.31
Average	**3.40**	**3,094**	**0.33**

living standards would be impossible. But these gains from depersonalized trade cannot be realized without a legal system that protects property rights and enforces contracts in an evenhanded manner. The failure of a country's legal system to perform these functions places a tight constraint on its prosperity.

The findings of the Economic Freedom Project are highly consistent with this view. The Legal System Area indicates the consistency of a nation's legal structure with protection of property rights, unbiased enforcement of contracts, independence of the judiciary, and rule-of-law principles. Among the approximately 100 countries for which data were available throughout the 1980–2000 period, twenty-four countries had an average Legal System Area rating of 7 or more. As Table 7 shows, these twenty-four countries had an average per capita GDP in 2000 of $25,716 and an average annual real growth rate of 2.5 percent over the two-decade period. Among these countries with a relatively sound legal system, the lowest 2000 per capita income levels were $11,960 and $13,279 for Hungary and Taiwan, respectively. The 2000 per capita GDP for twenty-two of the twenty-four countries exceeded $17,500. Perhaps even more important, all twenty-four of the countries with sound legal systems achieved positive real growth of per capita GDP over the two decades. In fact, only one of the twenty-four had an annual growth rate of less than 1.3 percent (Switzerland, at 0.82). Thus, all of the countries with sound legal systems grew and achieved relatively high levels of per capita income.

At the other end of the spectrum, there were twenty-one countries with an average 1980–2000 Legal System Area rating of less than 4. Among these countries, the average 2000 per capita GDP was $3,094 and the average growth rate during 1980–2000 was 0.33 percent. Both of these figures were approximately one-eighth of the parallel levels for the countries with sound legal systems. The highest 2000 per capita GDP among the twenty-one countries with a low-quality legal system was Colombia's $7,010. While five of these countries had growth rates of more than 2 percent, all of the five were exceedingly poor (per capita GDP of $3,400 or less). None of the twenty-one countries with low-quality legal systems was able to achieve both a 2000 per capita income of more than $3,400 and a growth rate during 1980–2000 of more than 1.1 percent. Thus, none of the countries with unsound legal systems was able to sustain a solid rate of growth once income levels rose above the $3,500 range!

All of this suggests that it will be virtually impossible for countries with legal systems that fail to protect property rights and enforce contracts to move up to even lower middle income status. These findings are also consistent with the view that countries lacking a legal system capable of enforcing contracts between parties who do not know each other and may well reside in different parts of the world will find it extremely difficult to achieve income levels that are the result of the gains derived from specialization, economies of scale, and depersonalized exchange.

CONCLUSION

During a debate with Milton Friedman about the Phillips curve and other aspects of stabilization policy during the late 1960s, Walter Heller attempted to refute Friedman's arguments by stating that he was an optimist. Friedman responded, "I am an empiricist." Milton Friedman has always been an empiricist, and that is why he was interested in developing a measure of economic freedom.

Some fifteen years ago, when I became involved with the project, I remember Milton Friedman stating that he was convinced that countries that were more economically free would grow more rapidly and achieve higher levels of income. However, he went on to note that without a measure of economic freedom, we were unable to directly test the validity of these hypotheses. The EFW measure now makes it possible for us to do so, and indeed, we have discovered that Friedman was right. But this is only part of the story. Critics often argue that a market economy leads to excessive income inequality, environmental degradation, extreme poverty, poor working conditions, and the like.[11] The EFW measure makes it possible for researchers to address these issues empirically. I believe that the major contribution of the EFW project will come from the use of the data as an empirical tool. Moreover, I'm sure that this contribution is one that will bring satisfaction to Milton Friedman.

NOTES

[1] For details on this series of conferences, see Walker (1988), Block (1991), and Easton and Walker (1992).

[2] The survey data are from the *International Country Risk Guide* (PRS Group 2001) and the *Global Competitiveness Report* (World Economic Forum 2003), two highly respected sources of information.

[3] Of course, a government may do a pretty good job in some areas, access to sound money and freedom of international exchange for example, and at the same time impose regulations that restrict economic freedom in other areas. The EFW area ratings can help identify the consistency of a country's policies with economic freedom in each of the five areas.

[4] Because some of the components are missing, particularly for years prior to 1990, a chain-link methodology similar to that used by national income accountants was employed to derive summary ratings for years prior to 2000. This chain-link methodology means that differences in a country's rating for years prior to 2000 will always reflect changes in the value of components available during overlapping years. This methodology enhances the comparability of country ratings across time periods.

[5] While the concept of economic freedom provides the compass for the design of the EFW index, the index can also be viewed in other ways. For example, some may perceive of the index as a measure of the quality of a country's institutional environment, a factor that has been stressed

in the writings of economists like Douglass North (1990), Peter Bauer (1957), Friedrich von Hayek (1960), Hernando de Soto (1989) and Gerald Scully (1988 and 1992).

[6] The EFW data were continuously available for 103 countries during 1980–2000. Because their per capita GDP figures and growth rates were dominated by conditions in the world market for crude oil, four of the countries (Bahrain, Kuwait, Oman, and the United Arab Emirates) are omitted from the analysis presented here.

[7] While the ratings for the 38 components of the EFW index are on a zero-to-ten scale, the range of the summary ratings is smaller. The mean summary rating for 1980–2000 for the ninety-nine countries of this study ranged from the 8.61 rating of Hong Kong to the 3.51 rating of the Democratic Republic of the Congo. The mean 1980–2000 summary rating for the ninety-nine countries was 5.69.

[8] The number of observations in the analysis of Table 4 is 91 (rather than 99) because the data on the growth of capital per worker (Kpw) were unavailable for eight countries. All of the omitted countries had a population of less than 1 million. The data on capital per worker are from Baier, Dwyer, and Tamura (2002).

[9] See Knack (1996) for additional evidence on the importance of institutions as a source of income convergence among nations.

[10] The following statement by Milton Friedman, made at the 2001 annual meeting of the Economic Freedom of the World Network held in San Francisco, highlights the importance of legal structure as a source of growth and prosperity:

> Just after the Berlin Wall fell and the Soviet Union collapsed, I used to be asked a lot: "What do these ex-communist states have to do in order to become market economies?" And I used to say: "You can describe that in three words: privatize, privatize, privatize." But, I was wrong. That wasn't enough. The example of Russia shows that. Russia privatized but in a way that created private monopolies—private centralized economic controls that replaced government's centralized controls. It turns out that the rule of law is probably more basic than privatization. Privatization is meaningless if you don't have the rule of law. What does it mean to privatize if you do not have security of property, if you can't use your property as you want to?

[11] For a survey article on economic freedom and a number of these topics, see Berggren (2003).

REFERENCES

Baier, Scott L., Gerald P. Dwyer Jr., and Robert Tamura. 2002. "How Important Are Capital and Total Factor Productivity for Economic Growth?" Federal Reserve Bank of Atlanta Working Paper Series no. 2002-2a (April).

Bauer, P. T. 1957. *Economic Analysis and Policy in Underdeveloped Countries*. Durham, NC: Duke University Press.

Berggren, Niclas. 2003. "The Benefits of Economic Freedom: A Survey." *The Independent Review* 8, 2 (Fall): 193–211.

Block, W., ed. 1991. *Economic Freedom: Toward a Theory of Measurement.* Vancouver, BC: The Fraser Institute.

De Soto, H. 1989. *The Other Path: The Invisible Revolution in the Third World.* New York: Harper & Row.

Dollar, David. 1992. *Exploiting the Advantages of Backwardness: The Importance of Education and Outward Orientation.* World Bank.

Easton, S. T., and M. A. Walker, eds. 1992. *Rating Global Economic Freedom.* Vancouver, BC: The Fraser Institute.

Gallup, John L., Jeffrey D. Sachs, and Andrew D. Mellinger. 1998. "Geography and Economic Development," in *Annual World Bank Conference on Development Economics 1998*, 127–70. Washington, DC: World Bank.

Gwartney, J., R. Lawson, and W. Block. 1996. *Economic Freedom of the World: 1975–1995.* Vancouver, BC: The Fraser Institute.

Gwartney, J. and R. Lawson, with Neil Emerick. 2003. *Economic Freedom of the World: 2003 Annual Report.* Vancouver, BC: The Fraser Institute.

Hayek, F. A. 1960. *The Constitution of Liberty.* Chicago: University of Chicago Press.

Knack, S. 1996. "Institutions and the Convergence Hypothesis: The Cross-National Evidence." *Public Choice* 87: 207–28.

North, D. 1990. *Institutions, Institutional Change, and Economic Performance.* Cambridge: Cambridge University Press.

Pritchett, Lant. 1997. "Divergence, Big Time." *Journal of Economic Perspectives* 11, 3 (Summer): 3–17.

The PRS Group. 2001. *International Country Risk Guide.* Syracuse, NY: The PRS Group Inc.

Scully, G. W. 1988. "The Institutional Framework and Economic Development." *Journal of Political Economy* 96, 3 (June): 652–62.

―――. 1992. *Constitutional Environments and Economic Growth.* Princeton, NJ: Princeton University Press.

Walker, M. A., ed. 1988. *Freedom, Democracy, and Economic Welfare.* Vancouver, BC: The Fraser Institute.

World Bank. 2003. *World Development Indicators.* CD-ROM.

World Economic Forum. 2003. *Global Competitiveness Report 2002–2003.* Oxford: Oxford University Press.

APPENDIX

The Areas and Components of the EFW Index

1. **Size of Government: Expenditures, Taxes, and Enterprises**
   A. General government consumption spending as a percentage of total consumption.
   B. Transfers and subsidies as a percentage of GDP.
   C. Government enterprises and investment as a percentage of GDP.
   D. Top marginal tax rate (and income threshold to which it applies).
      i. Top marginal income tax rate (and income threshold at which it applies).
      ii. Top marginal income and payroll tax rate (and income threshold at which it applies).

2. **Legal Structure and Security of Property Rights**
   A. Judicial independence: The judiciary is independent and not subject to interference by the government or parties in disputes.
   B. Impartial courts: A trusted legal framework exists for private businesses to challenge the legality of government actions or regulation.
   C. Protection of intellectual property.
   D. Military interference in rule of law and the political process.
   E. Integrity of the legal system.

3. **Access to Sound Money**
   A. Average annual growth of the money supply in the last five years minus average annual growth of real GDP in the last ten years.
   B. Standard inflation variability in the last five years.
   C. Recent inflation rate.
   D. Freedom to own foreign currency bank accounts domestically and abroad.

4. **Freedom to Exchange with Foreigners**
   A. Taxes on international trade.
      i. Revenue from taxes on international trade as a percentage of exports plus imports.
      ii. Mean tariff rate.
      iii. Standard deviation of tariff rates.
   B. Regulatory trade barriers.
      i. Hidden import barriers: no barriers other than published tariffs and quotas.
      ii. Costs of importing: The combined effect of import tariffs, license fees, bank fees, and the time required for administrative red tape raises costs of importing equipment by (10 = 10% or less; 0 = more than 50%).
   C. Actual size of trade sector compared with expected size.
   D. Difference between official exchange rate and black market rate.

E. International capital market controls.
   i. Access of citizens to foreign capital markets and foreign access to domestic capital markets.
   ii. Restrictions on the freedom of citizens to engage in capital market exchange with foreigners—index of capital controls among 13 IMF categories.

5. **Regulation of Credit, Labor, and Business**
   A. Credit market regulations.
      i. Ownership of banks: percentage of deposits held in privately owned banks.
      ii. Competition: Domestic banks face competition from foreign banks.
      iii. Extension of credit: percentage of credit extended to private sector.
      iv. Avoidance of interest rate controls and regulations that lead to negative real interest rates.
      v. Interest rate controls: Interest rate controls on bank deposits and/or loans are freely determined by the market.
   B. Labor market regulations.
      i. Impact of minimum wage: The minimum wage, set by law, has little impact on wages because it is too low or not obeyed.
      ii. Hiring and firing practices: Hiring and firing practices of companies are determined by private contract.
      iii. Share of labor force whose wages are set by centralized collective bargaining.
      iv. Unemployment benefits: The unemployment benefits system preserves the incentive to work.
      v. Use of conscripts to obtain military personnel.
   C. Business regulations.
      i. Price controls: extent to which businesses are free to set their own prices.
      ii. Administrative conditions and new businesses: Administrative procedures are an important obstacle to starting a new business.
      iii. Time with government bureaucracy: Senior management spends a substantial amount of time dealing with government bureaucracy.
      iv. Starting a new business: Starting a new business is generally easy.
      v. Irregular payments: Irregular, additional payments connected with import and export permits, business licenses, exchange controls, tax assessments, police protection, or loan applications are rare.

Variable Definitions and Sources

Variable Name	Variable Definitions	Variable Source
GDP per capita	GDP per capita (PPP US$)	World Bank (2003)
Annual growth rate of per capita GDP (1980–2000)	Average annual rate of growth in GDP per capita (real LCU) from 1980 to 2000	World Bank (2003)
I/GDP (1980–2000)	Average gross capital formation as a percentage of GDP from 1980 to 2000	World Bank (2003)
Investment per worker (1980–2000)	Average gross capital formation (US$) per worker from 1980 to 2000	World Bank (2003)
FDI per worker (1980–2000)	Average foreign direct investment (US$) per worker from 1980 to 2000	World Bank (2003)
EFW rating	Chain-weighted economic freedom rating	Gwartney and Lawson (2003)
Tropics	Percentage of population between Tropics of Cancer and Capricorn	Gallup, Sachs, and Mellinger (1998)
Coastal	Percentage of population within 50 km of coast	Gallup, Sachs, and Mellinger (1998)
Air distance	Distance by air from capital to New York, Rotterdam, or Tokyo (whichever is closest), 1,000s of km	Gallup, Sachs, and Mellinger (1998)
Kpw	Capital stock per worker (US$)	Baier, Dwyer, and Tamura (2002)
Hpw	Human capital stock per worker	Baier, Dwyer, and Tamura (2002)

Descriptive Statistics

Variable	N	Mean	Median	StDev	Min	Max
EFW rating, 1980–2000	99	5.69	5.57	1.09	3.51	8.61
Change in EFW rating, 1980–1990	99	0.50	0.57	0.65	−1.64	2.11
Change in EFW rating, 1990–2000	99	0.88	0.69	0.89	−0.69	4.04
GDP per capita, 2000 (US$)	99	10,669	6,033	10,564	490	50,061
GDP per capita, 1980 (US$)	99	4,263	2,551	3,825	362	14,534
Average annual growth rate of GDP per capita, 1980–2000	99	1.32	1.19	2.06	−4.58	8.15
I/GDP, 1980–2000	99	21.63	21.75	5.60	9.56	38.58
Investment per worker (US$), 1980–2000	99	3,813	1,727	4,942	48	21,142
FDI per worker (US$), 1980–2000	97	702	96	1,318	0	6,067
Coastal	99	0.54	0.57	0.38	0	1.00
Tropical	99	0.53	0.70	0.48	0	1.00
Air distance (km)	99	4,082	3,575	2,599	140	9,590
Kpw, 1999 (US$)	91	30,968	18,932	30,361	597	107,905
Hpw, 1999	91	6.43	6.21	2.16	2.60	11.34
Growth of Kpw, 1980–1999	91	1.97	1.66	2.56	−6.17	7.85
Growth of Hpw, 1980–1999	91	1.60	1.59	0.67	0.06	3.39

Can the Tide Turn?

Raghuram G. Rajan

Given that we are here to celebrate a work that is bereft of equations and charts, I thought I would try and emulate it in spirit by offering some reflections on whether the tide in favor of free markets can turn. If you are interested, however, I can point you to some empirical work that forms the basis of these reflections.

The second half of the last century seemed to have settled the debate over economic systems. Natural experiments like the one in Korea—where the South espoused capitalism and moved from underdeveloped to developed country status in a generation while the Socialist North descended into starvation and destitution—seemed to deliver a clear verdict: Capitalism is by far the best system for the production of wealth.

Yet, ironically, while capitalism has fattened people's wallets, it has made surprisingly few inroads into their hearts and minds. Many of the people taking to the streets against globalization are protesting against capitalism, which they accuse of oppressing workers, exploiting the poor, and making only the rich richer.

The tide can turn and markets can become shackled again. To prevent it from doing so, we have to understand why there was a wave in favor of free enterprise across the world toward the end of the twentieth century. We have to ask which of the then-favorable factors can change. And we have to use this understanding to prevent a reversal. Let us start by sketching why free enterprise has such a difficult time in so many parts of the world.

THE FORCES AGAINST ENTERPRISE

There is a growing consensus that for free enterprise to flourish, it needs a set of enabling institutions—a conducive legal environment, a decent supervisory and regulatory infrastructure, a reliable accounting system, etc.

In fact, recent work by Rodrik, Subramanium, and Trebbi (2002) suggestively entitled "Institutions Rule" shows that more than trade, more than geography, institutions seem to be the strongest factor in determining levels of per capita GDP. An increase in the level of institutions (as determined by an index for respect of property rights and an index for rule of law) from the level in Bolivia to the level in South Korea would be correlated with an increase in GDP per capita of 6.4 times, roughly the difference in GDP levels between the two countries.

The focus on institutions is not new; there is a long tradition dating back, at least, to Locke and Hume. What is refreshing, however, is that it is spilling over more and more into the policy debate and into popular work.

One example is Hernando de Soto's book *The Mystery of Capital* (2000). De Soto offers a simple explanation for why the poor do not have access to credit in developing countries. The answer, according to him, is that the poor do not have access to credit because they do not have official title to their properties. If slum dwellers could use the huts in which they live as collateral, they could borrow enough to set up small businesses and escape poverty.

The problem, according to de Soto, is that many of the poor are squatters and governments do not recognize their rights. Moreover, even if the poor have legitimate claims, obtaining title to property is prohibitively long and expensive. Finally, titles are often not enforced. Without well-enforced property rights, an arm's-length credit market cannot function and the poor become captive to moneylenders who exploit them.

While de Soto identifies an important problem, the lack of clear title (and, more generally, the fact that the poor have to work in an underworld without any formal structures supporting economic activity) is part of a greater problem. Even in countries where the poor have clear title, they are prevented from borrowing by laws that effectively discriminate against them.

For instance, in some U.S. states, individuals can file for bankruptcy and retain significant assets, including a home as a "homestead exemption." While such laws seem intended to benefit the poor, who would otherwise be homeless, a study by Gropp, Scholz, and White (1997) shows that they have the opposite effect. The law effectively deprives the poor of the ability to use their single biggest asset, their home, as collateral, since lenders know they can no longer seize the asset.

Moreover, asset-based finance is just one form of borrowing. There are others. For example, widespread information-sharing networks among creditors allow borrowers to access finance on the strength of their past credit histories. First-time borrowers can also get financing because they pledge their continued access to finance as collateral. If a borrower does not repay, lenders will share that information and ensure the borrower is cut off from future credit. This is sufficient deterrent that many borrowers continue paying, and knowing this, creditors are willing to lend. Many a small business has been started on the strength of unsecured credit card borrowings in developed countries.

That does not happen in many developing countries because information-sharing networks are underdeveloped. In fact, some countries have laws prohibiting the sharing of borrower information. Again, this has the effect of hurting the poor.

But even if we agree that underdeveloped institutions are the cause for disparities in income, there is the deeper question of why institutions are underdeveloped. At one level, the answer has to be found in politics because institutions are political. But why does politics play so differently across countries?

A hint to the answer comes from an influential World Bank study inspired by de Soto's early work. It finds that in Peru, the number of days it takes to satisfy all the permissions needed to start a new business is over 171. By contrast, the number of days it takes to obtain all the permissions to start a new business in Canada is two, and in the United States it is seven. In our own work, we have found that countries where financial markets are underdeveloped (and thus access to credit restricted) are also countries where the number of days to start a new business is high. In other words, the lack of access to credit is a barrier to entry much like the permissions needed to open businesses.

There is a deeper and broader pattern here than simply the lack of recognition of customary property rights—there is a concerted effort to limit widespread access (and not just for the poor) to markets, either by actively creating impediments or by leaving the necessary infrastructure underdeveloped. It is necessary to understand why this pattern exists in order to propose solutions.

Our Explanation

Our explanation is simple. In many countries, powerful elites oppose widespread access to markets. They have the political clout to erect direct impediments like mandatory permits to open a business or indirect barriers like an inadequate infrastructure. The reason for their opposition is obvious. These elites already get what they need from the limited markets that exist. They stand to lose if access to markets became freer and they faced competition. As a result, ordinary people never see true free market capitalism, which implies competition and equal access. They only experience the failed version, which destroys hope. Unfortunately, both systems get tarred with the same brush, and capitalism is seen as a system of the rich, by the rich, and for the rich.

All we are suggesting is a general tendency, not an ironclad law: When markets start out limited, those who already have access often have very different incentives from those who don't. If, as is likely, the former are more politically powerful because of the wealth they obtain from their privileged access to markets, they can ensure the government does not create the conditions for wider access. Of course, they may not need to campaign actively against market-friendly infrastructure. The government may simply not be interested in the welfare of the commoner, and the politically influential and conscious classes may not see fit to change matters. Neglect can be as effective as overt opposition.

Of course, this begs the question of why initial conditions are such that only some have access. One answer may lie in historical origins. An intriguing study by Acemoglu, Johnson, and Robinson (2001) starts with the idea that the propensity of European settlers to settle in areas they colonized is negatively correlated to the extent of settler mortality rates in those areas. Not surprisingly, Europeans did not flock to areas where yellow fever and malaria were endemic, preferring to send a few overseers who could keep the more adapted locals in check. Governance was more exploitative and hierarchical in those colonies and created a small group of elite ruling over a much larger group of natives. Acemoglu, Johnson, and Robinson find a correspondence between the current quality of institutions and settler mortality. The channel may be because there was a natural creation of elites in those countries.

Similarly, one could argue that even if settler mortality in parts was low, a large local population would make it hard for many new settlers to enter. So the colonization of a country like India had to proceed through the creation of local elites who would owe allegiance to the colonizer. These elites would continue to dominate long after the departure of the colonists and the coming of independence and democracy. Gandhi's greatest fear was that the white sahib would give way to the brown sahib, and in many countries that fear has been realized.

Let us summarize our arguments. Our point is that the absence of infrastructure supporting markets in much of the world is not because developing countries do not know that well-defined property rights and transparent contracting environments are of vital importance. In only a very few countries is it because the country does not have the physical or human capital to build infrastructure. In many more, it is because there are too many interests against the building of the right infrastructure.

No wonder, then, that the poor around the world see markets as being against them, not realizing that what they experience is a very corrupted version of the markets we experience. That the elite tend to have disproportionate say in the running of markets is actually a point on which the right and the left agree—one of the few places where Chicago economist George Stigler echoes Karl Marx. But the left is wrong in saying that markets need to be replaced by the government, for that will just perpetuate the capture by the elite.

And the extreme right is wrong in saying we can dispense with the government: The absence of government can also be anticompetitive and retard free markets.

Consider another example. If you wanted to fly and there were no supervisory authority in the airline industry and no regulations enforcing safety standards, you would be very reluctant to fly fledgling airlines. You would prefer the established ones that had the track record and the reputation. So a complete lack of safety regulations in the airline industry would favor established firms, making entry impossible and killing competition.

What we need, therefore, is a Goldilocks government, not too interventionist and not too laissez-faire, a government that is just right. The difficulty with this Goldilocks position is, of course, in implementation. If we recognize that the government is controlled by special interests, how can we hope it will create just the necessary infrastructure for wider markets without constantly interfering in their working? It is naïve to assume that money will stop mattering in politics, no matter how much electoral reform takes place. The challenge then is to get the elites behind markets rather than against them, for them to see that opportunities from expanded markets outweigh the increased competition.

HOW DID THE TIDE TURN?

Somehow, this has happened in the last three decades. Markets have been spreading. Why have politicians in countries as diverse as France and Germany or Korea and India embraced the market and attempted to provide the governance markets need? It is difficult to imagine that politicians have suddenly become more public spirited. The answer, we believe, is that the interests of the elites have changed with the opening of borders to goods and capital. This has made domestic elites press their politicians to enact market-friendly legislation.

The effect of open borders can be clearly seen in the Indian automobile industry. In the mid-1980s, the industry was protected from foreign imports. The result was that consumers had a choice between just three car models, and only one if they wanted a big car. The cars were obsolete gas-guzzlers, unchanged in design for decades. (The biggest car, the Ambassador, had been designed in England nearly forty years before.) Nevertheless, the car-starved public was willing to wait for years to be allocated one of these monstrosities. The rationale for not allowing foreign producers into India was, in part, that the domestic manufacturers would be wiped out if the market were opened up, and in part that cars were frivolous luxuries and there were more important goods for consumers to spend on. The truth was that producers were being pampered at the expense of the consuming public.

In the early 1990s, following a financial crisis, the Indian government opened the car market to foreign producers. The worst fears of the domestic producers were realized. The public abandoned them for the new foreign models, and the old manufacturers were wiped out (though since then the Ambassador has rediscovered a market as a "period" car in the West). But it simply did not make sense for the foreign manufacturers to continue sourcing their subassemblies from outside India. Instead, they started developing local auto ancillary manufacturers and gave them the technological assistance to become world-class. Soon India started exporting ancillary automotive products to the developed world.

The story does not end here. An Indian manufacturer, Telco, capitalizing on the existence of world-class suppliers of ancillaries in India, started producing a state-of-the-art, indigenously designed car, the Indica. The car had teething problems at first and was rejected by a now-discriminating public. But Telco engineers went back to the drawing board, fixed the flaws, and brought out a new version that swept the market in its category. From about 50,000 cars in the early 1980s, India now produces over 600,000. Next year, it is slated to export 200,000 cars, many to the developed world. The automobile ancillary industry grew by 20 percent in sales last year and by an average of about 10 percent in the decade before that. The Indian automobile industry offers an example of what trade liberalization can bring—potentially some pain in the short run, but enormous gain in the long run.

In sum, as borders open up to the flow of goods and capital, incumbent firms now need well-functioning domestic markets so they can take advantage of the opportunities provided by the global market, as well as meet foreign competition head-on. The prospect of increased domestic competition matters less when firms are fighting on the world stage. They now back, rather than oppose, domestic markets. Put differently, competition between economies through open borders forces politicians to enact the rules that will make their economies competitive. This typically means enacting market-friendly legislation and making markets accessible to all.

This is not to say that open borders force a bland uniformity on everyone, a golden straitjacket, in the words of Thomas Friedman. Instead, they persuade governments to find the best path for their own peoples. In some countries, that might mean sixty-hour workweeks with high pay and low benefits; in others, it might mean thirty-five-hour weeks with lower pay and lots of benefits. The point, however, is that the package of work, productivity, and pay has to be competitive, which means meaningless and harmful rules have to wither away.

CAN THE TIDE TURN?

We have suggested that a primary factor in the growth of domestic markets and free enterprise across the world was the progressive opening up of borders. Clearly, other factors were also at work.

In particular, the ideas of Milton Friedman, Friedrich Hayek, George Stigler, and others of the Chicago school of economics offered a respectable alternative to Keynesian economics, which held sway in the higher echelons of economics after World War II. And unlike an increasing number of their academic confreres, these economists did not shrink from addressing the public directly. Milton and Rose Friedman's *Free to Choose* and Friedrich Hayek's *Road to Serfdom* were popular successes (the latter after selling slowly initially) and did much to persuade generations of young people of the perils of excessive government.

Despite the widespread impact of these ideas, it would be too simplistic to couch the tide toward free enterprise toward the end of the twentieth century as simply the consequence of politicians and the public becoming convinced of their truth. If nothing else, the timing is not right. Hayek wrote his searing critique of the managed economy, *The Road to Serfdom*, in 1944, but it was only with Margaret Thatcher's accession to power in 1979 that a major government was willing to espouse his ideas. The Friedmans' oeuvre was closer in time to the presidency of Ronald Reagan, but it reflected a lifetime of ideas for which Milton Friedman had already won a Nobel Prize.

Instead, it is better to think of changes in economic attitudes as a consequence of the fortuitous combination of ideas, events, and interests, with each playing its own part. The stagflation experienced by developed economies in the 1970s was important in forcing economists to look to new ideas. With the breakdown of the Bretton Woods system, the United States, pressed by its bankers, put its weight behind making borders as open to the flow of capital as they were becoming open to the flow of goods. In turn, this unleashed competitive forces that gave domestic governments and business groups strong incentives to improve the competitiveness of their economies. That the tide was not just driven by ideas is suggested by the fact that the Socialist government of Francois Mitterand turned in the span of a few years from nationalizing enterprises to privatizing them.

This is not to say that ideas do not matter—the existence of a coherent and respectable alternative to the Keynesians in Chicago was critical in giving sympathetic politicians ammunition and in persuading the larger public. But to think that ideas are all that matter is to foster dangerous complacency. For if the expansion of the free enterprise system did not solely reflect the supremacy of the ideas behind it, those ideas may not be enough to preserve its position. Interests and events may, unfortunately, now be moving in less propitious ways. Start first with interests.

Interests

We have argued that open borders have been instrumental in increasing competition. This has forced domestic elites to create market infrastructure and expand access to goods markets and finance to everyone. Of course, the decision to open borders is itself political. It has been an easy one in recent years because the largest economies in the world, foremost among them the United

States, have become more open. Not only has this provided more opportunities for other countries that open up, it also has made it harder for countries to remain closed because with the world largely open, goods and capital can leak easily across a closed economy's borders. However, this means that open borders are hostage to the intentions of the largest economies. If they become protectionist and turn inward, smaller economies will follow. The closing of borders will weaken pro-market interests and strengthen anti-market ones.

Open borders are especially under risk in times of downturn, when the foreigner and his goods become an easy political target. But recent developments may make developed countries less willing champions of open borders: Typically, as developing countries have been persuaded to open their markets in return for developed country market access, there has been a stable political outcome. The developed countries take the high-skill, high-knowledge, high-capital-intensity end, while developing countries take the low end. Eventually, developing countries move up, but in the initial phase the high-intensity industries in developed countries, which are typically more politically powerful in their own countries, see gains and push for more openness. The result has been a strong impetus for freer trade.

Survey evidence shows that in the past, workers in nontraded industries used to be strong supporters of free trade. The big change now is that many sectors that used to be nontraded are now becoming traded. An accountant in the Philippines can now do your taxes via the Internet as effectively as someone from H&R Block across the street. The service sector in developed countries is especially unaccustomed to such competition and can react forcefully. This may alter the balance of forces for free trade, even in developed countries. Already some states in the United States are threatening to blacklist businesses that outsource processes to other countries.

The attitude of the world's largest economy is particularly important in determining whether protectionism will come to dominate the world over. And this country has sent mixed signals in the recent past. The steel tariffs were followed by enhanced farm subsidies. Politicians are entering the fray with complaints about unfair trade practices in China. More restrictions may be on the way.

Trade protectionism is particularly detrimental to capitalism elsewhere because it undercuts strong constituencies for free markets in other countries and prevents those borders from opening up fully. And if unchecked, other countries will take an eye for an eye, making the whole world eventually go blind.

Events

Corporate scandals have not helped. One of the reasons people in this country tolerate the enormous inequalities in income and wealth, say, relative to Europe, is because there is a perception that markets are fair and success is accessible to all. Corporate scandals suggest that the CEO, his investment banker, and their friends play on a different playing field than the ordinary pensioner. The danger is that ordinary people might get disillusioned with reform and shift their support to demagogues who seek to sandbag the market. Equally bad would be if the reforms get hijacked by special interests that use the turmoil to carve out their own little privileges. Sarbanes–Oxley, intended to strengthen corporate governance, is viewed by some as the Big Accounting Firm Profit Restoration and Protection Act of 2002 because of the central role it implicitly accords to the Big Four accounting firms.

More generally, these scandals have given the free enterprise system a bad name across the world. The actions of one company, Enron, have done more to persuade the modern youth of capitalism's evils than the myriad unread(able) texts of the doctrinaire radical left.

In sum, interests and events are coalescing again but, unfortunately, in the wrong direction. As developed country governments become less sure of the cause of free markets both from the perspective of their own interests and as a moral imperative, less economically sophisticated groups take over the debate. Emotions now prevail over evidence. Misguided activists persuade developing country governments that their instinctive mercantilism and opposition to free trade are not just economically sound but also the right response for their own people. With no one championing the cause of free trade, debacles like Cancun are the unfortunate consequence.

With events and interests not cooperating, it is important that we turn back to ideas to strengthen us and to prevent the tide from turning. We need to marshal the wealth of evidence we have on the benefits of trade and engage dissident economists and demagogic activists in fruitful dialogue, instead of letting them dominate the public arena. We need to persuade the public that corporate scandals are aberrations that can be fixed rather than the norm in a system of free enterprise. We need to combat the drift in these dangerous times of reposing our faith in the government in all arenas.

But we have an example of how this can be done. Milton Friedman's life offers us a guiding beacon on how ideas can be used to hold back the power of events and interests. Let us learn from him, because the battle needs to be fought over and over again.

CONCLUSION

The developed world needs markets to grow in the developing world, not just because unrest in the latter leads to swarms of economic refugees into the former, nor just because aging, retired populations in developed countries will have to depend increasingly on productive young populations in developing countries, nor even because those markets will absorb developed country goods today. The developed world needs a more open, competitive developing world because that is the best guarantee that capitalism in the developed world will stay vibrant, with capitalists motivated more by the opportunities for growth than by the fear of competition. We cannot let the tide turn. Instead we have to continue to break the shackles holding back markets the world over. On that depends our own future.

REFERENCES

Acemoglu, Daron, Simon H. Johnson, and James A. Robinson. 2001. "The Colonial Origins of Comparative Development: An Empirical Investigation." *American Economic Review* 91: 1369–1401.

De Soto, Hernando. 2000. *The Mystery of Capital: Why Capitalism Triumphs in the West and Fails Everywhere Else*. New York: Basic Books.

Gropp, Reint, John Karl Scholz, and Michelle J. White. 1997. "Personal Bankruptcy and Credit Supply and Demand." *Quarterly Journal of Economics* 112: 217-51.

Rodrik, Dani, Arvind Subramanian, and Francesco Trebbi. 2002. "Institutions Rule: The Primacy of Institutions over Geography and Integration in Economic Development." CEPR Discussion Paper No. 3643 (November).

Acknowledgments

The papers in this volume were presented at a conference held at the Federal Reserve Bank of Dallas in October 2003. We wish to thank the authors as well as Milton and Rose Friedman for writing the foreword and Alan Greenspan for allowing publication of his remarks. (Two of the conference speakers, Gary Becker and Francisco Gil Diaz, are not represented in these proceedings.) We also extend special thanks to Kay Gribbin and Betty Chapman of the Research Department and Gloria Brown, Laurel Brewster, and Heather McDonald of the Public Affairs Department for assistance in organizing the conference. We would also like to thank Kay Champagne, Jennifer Afflerbach and Laura Bell of the publications staff for seeing this volume through to publication.

MARK WYNNE
Vice President and Senior Economist
HARVEY ROSENBLUM
Senior Vice President
and Director of Research
ROBERT L. FORMAINI
Senior Economist

Federal Reserve Bank of Dallas
Dallas, Texas

About the Contributors

Terry L. Anderson is a senior fellow at the Hoover Institution at Stanford University, professor emeritus at Montana State University, and the executive director of PERC—The Center for Free Market Environmentalism. Anderson helped launch the idea of "free market environmentalism" with the publication of his book by that title, coauthored with Donald Leal. Anderson is the author or editor of 28 books.

Ben S. Bernanke is a member of the Board of Governors of the Federal Reserve System. Before joining the Board, he was chair of the economics department at Princeton University. Bernanke has also been a visiting scholar at the Federal Reserve Banks of Philadelphia, Boston, and New York. He has published numerous articles on a variety of economic issues, including monetary policy and macroeconomics, and has authored several books.

Peter J. Boettke is deputy director of the James M. Buchanan Center for Political Economy, senior research fellow at the Mercatus Center, and economics professor at George Mason University. He is the author of three books dealing with the history, collapse, and transition from socialism, and he has recently become a coauthor on Paul Heyne's classic textbook, *The Economic Way of Thinking*. Before joining the faculty at George Mason, Boettke taught in the economics department at New York University and was a national fellow at the Hoover Institution.

Gregory C. Chow has authored more than 170 articles and 11 books, including *China's Economic Transformation*. The Chow test for stability of parameters is found in every econometrics textbook. He advised top government officials in Taiwan and mainland China on economic policy and economic reform and worked with the Chinese Ministry of Education to introduce the teaching of modern economics there. He has held academic posts at MIT, Cornell University, and Columbia University. In 1970, Chow became director of the Econometric Research Program at Princeton University, which was renamed the Gregory C. Chow Econometric Research Program when he retired in 2001.

Tyler Cowen is the Holbert L. Harris University Professor of Economics at George Mason University and general director of both the James M. Buchanan Center and the Mercatus Center. He is author of *In Praise of Commercial Culture*, on the economics of music and the arts, and *What Price Fame?* on the economics of fame. He cowrote *Explorations in the New Monetary Economics* with Randall Kroszner. Cowen recently published *Creative Destruction: How Globalization Is Changing the World's Cultures* on the economics of multiculturalism.

James D. Gwartney is professor of economics and director of the Gus A. Stavros Center for Free Enterprise and Economic Education at Florida State University. In 1999–2000 he served as chief economist of the Joint Economic Committee of the U.S. Congress. Gwartney has authored or coauthored numerous books and articles, including the annual *Economic Freedom of the World* report and *Economics: Private and Public Choice*, a textbook used by more than a million students over the past two decades.

Eric A. Hanushek is a senior fellow at the Hoover Institution at Stanford University, a member of the Koret Task Force on K–12 Education, and a research associate at the National Bureau of Economic Research. He is chairman of the executive board of the Texas Schools Project at the University of Texas at Dallas. Hanushek is a leading expert on educational policy, specializing in the economics and finance of schools, and has authored nine books on the topic.

Laura E. Huggins is a research fellow at Stanford University's Hoover Institution, where she specializes in free market environmentalism, property rights, and population policy. Her primary interest is economic processes' role in shaping natural resource policy and promoting market principles to help resolve environmental dilemmas. Huggins coauthored *Property Rights: A Practical Guide to Freedom and Prosperity* and *The Property Rights Path to Sustainable Development* with Hoover senior fellow Terry Anderson.

Allan H. Meltzer is professor of political economy at Carnegie Mellon University and a visiting scholar at the American Enterprise Institute. He has served as a consultant on economic policy for Congress, the U.S. Treasury, the Federal Reserve, the World Bank, and foreign governments. His work in the history of U.S. monetary policy, government size, macroeconomics, and international financial reform earned him the Distinguished Fellow award from the American

Economic Association in 2002 and the Lifetime Achievement Award from Money Marketeers, New York University, in 1997.

William A. Niskanen, a former defense analyst, business economist, and professor, has been chairman of the Cato Institute since stepping down in 1985 as acting chairman of President Reagan's Council of Economic Advisers. Niskanen specializes in the areas of policy analysis and public choice. Before serving four years at the Council of Economic Advisers, he was director of economics at Ford Motor Co. for five years and assistant director of the Office of Management and Budget for two years. His latest book is *Autocratic, Democratic, and Optimal Government: Fiscal Choices and Economic Outcomes*.

Paul E. Peterson is Henry Lee Shattuck Professor of Government at Harvard University, director of Harvard's Program on Education Policy and Governance, a senior fellow at the Hoover Institution, and editor in chief of *Education Next: A Journal of Opinion and Research*. He is coauthor of *The Education Gap: Vouchers and Urban Schools*. His work earned him an appointment to a Department of Education independent review panel evaluating the Title I program for disadvantaged students.

Raghuram G. Rajan is the economic counselor and director of research at the International Monetary Fund. He is on leave from the Graduate School of Business at the University of Chicago, where he is the Joseph L. Gidwitz Professor of Finance. His research in corporate finance, theory of organizations, and financial intermediation and regulation has led him to visiting professorships at MIT, Northwestern University, and the Stockholm School of Economics. He and Luigi Zingales are authors of the book *Saving Capitalism from the Capitalists*.

Thomas R. Saving is director of the Private Enterprise Research Center and University Distinguished Professor of Economics at Texas A&M University. His research has covered antitrust economics, monetary economics, the theory of the banking firm, and the general theory of firms and markets. In 2000, President Clinton appointed him to the Board of Trustees of the Social Security and Medicare Trust Funds. In 2001, President Bush appointed him to the President's Commission to Strengthen Social Security.

Richard L. Stroup is a senior associate at PERC—The Center for Free Market Environmentalism. He is also professor and interim head of the Department of Agricultural Economics and Economics at Montana State University. Stroup has served as director of the Interior Department's Office of Policy Analysis and published widely in professional journals, trade books, and popular media publications on the economics of resources and the environment. His most recent book is *Eco-Nomics: What Everyone Should Know about Economics and the Environment*.

Luigi Zingales is the Robert C. McCormack Professor of Entrepreneurship and Finance at the University of Chicago. He is also a fellow at the National Bureau of Economic Research, the Centre for Economic Policy Research, and the European Corporate Governance Institute. Zingales coauthored *Saving Capitalism from the Capitalists* with Raghuram Rajan. He has had more than two dozen publications in such professional journals as *American Economic Review*, *Quarterly Journal of Economics* and *Journal of Finance*.